Plato and the Nerd

Plato and the Nerd

The Creative Partnership of
Humans and Technology

Edward Ashford Lee

The MIT Press
Cambridge, Massachusetts
London, England

This book was set in Times Roman using LATEX. Printed and bound in the United States of America.

A note about the cover design: The cover takes its inspiration from M. C. Escher's 1948 lithograph "Drawing Hands," which showed two hands in the act of drawing each other.

Library of Congress Cataloging-in-Publication Data

Names: Lee, Edward A., 1957- author.
Title: Plato and the nerd : the creative partnership of humans and technology / Edward
    Ashford Lee.
Description: Cambridge, MA : The MIT Press, [2017] | Includes
    bibliographical references and index.
Identifiers: LCCN 2016056891 | ISBN 9780262036481 (hardcover : alk. paper)
Subjects: LCSH: Technology–Philosophy. | Computer
    science–Popular works. | Creative ability.
Classification: LCC T14 .L4254 2017 | DDC 601–dc23 LC record
    available at https://lccn.loc.gov/2016056891

10   9   8   7   6   5   4   3   2   1

This book is dedicated to my muse, Rhonda Righter, with thanks for many dinnertime conversations that shaped my thinking.

Contents

Preface

What This Book Is About

When I was young, my father wanted me to become a lawyer or get an MBA and take over the family business. Engineers were the people who worked for him. The brightest young minds, at least those of white Anglo-Saxon stock in the United States, went to law school, business school, or medical school. Today, engineering schools are much harder to get into, but that was not true when I was going to college. Yes, my father was profoundly disappointed in me when I double majored in Computer Science and what Yale called "Engineering and Applied Science." I made it worse when I went to MIT for graduate school in engineering and then went to work as an engineer at Bell Labs, and worse still when I went to Berkeley for a PhD and then became a professor. This book is perhaps my last-ditch attempt to justify those decisions.

When I started writing the book, I really didn't know who my target audience would be. As it has turned out, this book is targeted toward readers who are either literate technologists or numerate humanists. I'm not sure how many such people there are, but I'm convinced there must be a few. I hope you are one of them.

This book is my attempt to explain why the process of creating technology, a process that we call engineering, is a deeply creative process, and how this explains why it has become so hot and competitive, making geeks out of the brightest young minds. The book is about the culture of technology, about both its power *and* its limitations, and about how the real power of technology stems from its partnership with humans. I like to think of the book as a popular philosophy of technology, but I doubt it will be very popular, and I am not sure I have the qualifications to write about philosophy. So really, the only guarantee I can make is that it is about technology and the engineers who create technology. And even then, it is limited to the part of technology that I understand best, specifically, the digital and information technology revolutions.

This book is *not* about the artistry and creativity that is unleashed by using technology as a medium. For that topic, I recommend the wonderful book by Virginia Heffernan, *Magic and Loss* (Heffernan, 2016). Heffernan claims that "the Internet is a massive and collaborative work of realist art," but she is referring to the *content* of the Internet. In my book, I claim that Internet *technology itself*, and all of digital technology that shores it up, is a massive and collaborative creative work, even if not an artistic work.

Digital technology as a medium for this latter sort of creativity has enormous potential, well beyond what has been accomplished to date. In the first part of this book, I explain exactly why this technology has been so transformative and liberating. I study how engineers use models and abstractions to build inventive artificial worlds and give us incredible capabilities, such as the ability to carry around in our pockets everything humans have ever published.

But this is not to say that digital technology has no limitations. Pursuing a yin and yang balance, in the second part of the book, I attempt to counter a runaway enthusiasm among some thought leaders about digital technology and computation. Driven by the immense potential of computers, this enthusiasm has led to unjustified beliefs that go as far as to assert that everything in the physical world is in fact a computation, in exactly the same sense as in modern computers. Everything, including such complex phenomena as human cognition and such unfamiliar objects as quasars, is software operating on digital data. I will argue that the evidence for such conclusions is weak and the likelihood is remote that nature has limited itself to only processes that conform with today's notion of digital computation. And I will show that this digital hypothesis cannot be tested empirically, and therefore can never be construed as a scientific theory. Because the likelihood is remote, the evidence is weak, and the hypothesis is untestable, these conclusions are an act of faith. My argument here will likely get me into trouble because I'm swimming against a considerable current.

Also bucking much current thought, I argue that the goal of artificial intelligence to reproduce human cognitive functions in computers is misguided, is unlikely to succeed, and vastly underestimates the potential of computers. Instead, technology is coevolving with humans, augmenting our own cognitive and physical capabilities, all the while enabling us to nurture, evolve, and propagate the technology. We are seeing the emergence of symbiotic coevolution, where the complementarity between humans and machines dominates over their competition.

But most of the book is very much swimming with the current, upbeat about the enormous potential of technology to improve our lives. But more than just utilitarian, one of my main messages is that engineering is a deeply creative and intellectual discipline, every bit as interesting and rewarding as the arts and sciences. In areas where the technology is less mature, the creative contributions reflect the personalities, aesthetics, and idiosyncrasies of the creators. In areas that are more mature, the work can become deeply technical and opaque to outsiders. This happens in all disciplines, so this is hardly surprising.

Like the sciences, engineering is built around accepted paradigms that provide frameworks for thought. Also like the sciences, engineering is punctuated by paradigm shifts, to use the words of Thomas Kuhn (Kuhn, 1962). Unlike the sciences, however, the paradigm shifts are frequent, even relentless. I argue, in fact, that the pace of technological progress in our current culture is more limited by our human inability to assimilate new paradigms than by any physical limitations of the technology. I attempt in this book to explain why this is.

Like the arts, the evolution of the field of engineering is governed by culture, language, and cross-germination of ideas. Also like the arts, success or failure is often determined by intangible and inexplicable forces, such as fashion and culture. And in an observation that may take many readers by surprise, also like the arts, the creative media used to engineer new artifacts and systems today, particularly digital media, have become astonishingly versatile and expressive. In my opinion, this latter property, the versatility and expressiveness of digital media, accounts for the attractiveness of the field to bright young minds, more even than the lucrative job prospects.

Engineering is a broad field, encompassing everything from water supply systems to social networking software. Any individual, myself included, cannot have more than a superficial understanding of more than a few of its subdisciplines. My arguments in this book, therefore, are based on my experience with electronics, electrical engineering, and computer science. These arguments apply to digital and information technologies and may or may not apply to other technologies such as bridges and chemical plants. Nevertheless, I do know from experience that digital technologies have invaded nearly all other engineering disciplines. Modern chemical plants, for example, include substantial computer control and therefore become instances of cyber physical systems, discussed in chapter 6. Such systems are most certainly subject to the potential, vagaries, and limitations of digital technology that I point out in this book.

I do not assume of the reader any particular technical background. In some sections of the book, I do dive more deeply than I probably should into technical topics that are near and dear to my heart, but I promise the reader that every such indulgence is short, and hopefully skipping the technical details will not seriously undermine the message. Please persist. The nerd storm will pass quickly.

I do assume a numerate reader. Against all advice, I have even included 12 equations in the book. They are not complicated equations. High school math and science is more than sufficient to fully understand them, but even then full understanding is not needed to get the message. My publisher has used this argument against me, saying that if it is true, I should remove them. But I like them. I have confidence that there are more numerate readers than there used to be. I have assured the publisher that, counting my friends and family, a few dozen book sales are assured.

The title of this book comes from the wonderful book by Nassim Nicholas Taleb, *The Black Swan* (Taleb, 2010), who titled a section of the prologue "Plato and the Nerd." Taleb talks about "Platonicity" as "the desire to cut reality into crisp shapes." Taleb laments the ensuing specialization and points out that such specialization blinds us to extraordinary events, which he calls "black swans." Following Taleb, a theme of my book is that technical disciplines are also vulnerable to excessive specialization; each speciality unwittingly adopts paradigms that turn the speciality into a slow-moving culture that resists rather than promotes innovation.

But more fundamentally, the title puts into opposition the notion that knowledge, and hence technology, consists of Platonic Ideals that exist independent of humans and is discovered by humans, and an opposing notion that humans create rather than discover knowledge and technology. The nerd in the title is a creative force, subjective and even quirky, and not an objective miner of preexisting truths.

I hope that through this book, I can change the public discourse so young people are more inclined to consider a career in engineering, and not just because of the job prospects. I am convinced that engineering is fundamentally a creative discipline, and the technical drudgery that prejudices many people is no more drudgery than found in any other creative discipline. Yes, hard work is required, but as a reward for that hard work, you can change the world.

Overview of the Chapters

Some readers like to be told what they will be told before they are told it. Putting aside the problematic self-referentiality, for those readers, I provide here a brief overview of the book. But honestly, I recommend skipping this and going directly to chapter 1. The story told in this book cannot be accurately summarized in a few paragraphs, and any such summary will necessarily make the book seem more dense than it is. Nevertheless, for those who really need it, here is my summary.

Popular perception of technology and engineering is often one of a dispassionate field dominated by logic and trading in colorless facts and truths. In chapter 1, I explore the idea of facts and truths in technology, showing that these are not just discovered but more often invented or designed. Rather than being built on timeless Platonic Ideals, technology is built on ideas that are more fluid and sometimes quirky. The notion of truth becomes more subjective; collective wisdom becomes better than individual wisdom; a narrative about how facts evolve becomes more interesting than the facts themselves; facts and truths may be wrong; and it can cost billions to show that facts are true. I then develop the idea that engineering and science, disciplines rooted in facts and truths, are complementary and overlapping, leveraging each others' methodologies. In this chapter, I try to understand the cultural phenomenon that engineering has been considered the "kid sister" of science.

In chapter 2, I focus on the relationship between discovery and invention. A key theme of this chapter is that models are invented not discovered, and it is the usefulness of models, not their truth, that gives them value. Note that the usefulness of a model need not be a practical, utilitarian sort of usefulness. A model may be useful simply because it explains or predicts observations, even if the phenomena observed have no practical application.

Models are useful to scientists when they are faithful to the natural system being studied, whereas models are useful to engineers when a physical realization can be constructed that is faithful to the model. These uses are complementary and, in fact, are often applied in combination.

Chapter 2 is heavily influenced by Kuhn (1962). But Kuhn focused on science, not engineering. The engineering use of models results in more room for creativity in the construction of models because it is not necessary for the models to be faithful to some preexisting natural system. But the use of models can also slow technological change because models are built on paradigms that frame our thinking and therefore limit our thinking. Models can also get quite sophisticated, forcing increased specialization, which can also slow change by impeding communication across specializations.

In chapter 3, I dive into exactly how the engineering use of models enables creativity. I do this by illustrating the role that models have played in the development of digital technology, where models are stacked many layers deep, with the design of each layer affecting the designs both above and below it. Digital technology has, through this multiplicity of layers, mostly removed any meaningful physical constraints from a broad class of engineered systems. Each layer of models conforms with an established paradigm, a way of modeling and abstracting an engineered design. Innovation, therefore, is less limited by the physics of the technology than by our imagination and ability to assimilate new paradigms.

I argue that paradigms play a central role in digital technology because without them, no human could possibly comprehend the complexity of the systems we routinely build today. But these paradigms are human constructions, governed by culture and language. In many cases, the paradigms that have emerged are idiosyncratic, reflecting the personality and aesthetics of their creators.

A notable feature of digital technology is that paradigms are layered one on top of another. Semiconductor physics gives us the ability to make transistors, which we can use as electrically controlled switches that have two distinct states: "on" and "off." This enables a digital abstraction that turns out to be just the first of many layers, building up eventually to the programming languages that enable us to build databases, machine learning systems, web servers, and so on. Each of these layers forms through coalescing of competing paradigms.

In chapter 4, I explore the layered paradigms that make up much of today's digital technology hardware. I show that the physical substance of the hardware is not durable, but the paradigms are. The hardware is routinely discarded every few years as it wears out and becomes obsolete, but the principles on which the hardware is designed, with all their warts and idiosyncrasies, persist for decades.

In chapter 5, I explore the layered paradigms that make up much of today's *information* technology. These paradigms define how we construct software, and software, it turns out, endures much better than hardware. Paradigms, like human culture, change slowly, particularly compared with the speed with which technology changes. Although Kuhn's scientific paradigms are strictly human constructions, the paradigms of software are encoded in the software. In an orgy of self-referentiality, software builds its own scaffolding. The self-scaffolding of software makes it much more durable than hardware, despite its ephemeral nonsubstantive existence. It could even outlast humans.

Chapter 6 explores the structure of technology revolutions, with a particular focus on digital technology. This chapter is also heavily influenced by Kuhn, but it strives to identify how technology revolutions differ from scientific revolutions. One key difference is that technology paradigms appear and disappear much more rapidly probably because, compared with scientific paradigms, they are relatively unconstrained by the physical world and are layered one upon another many layers deep. Like scientific paradigms, new technology paradigms do not necessarily replace old ones. They may instead overlay the old ones, building new platforms on top of existing platforms. The ability to do this depends on the transitivity of models explored in the three previous chapters. Unlike scientific paradigms, the crises that trigger new technology paradigms do not arise so much from the discovery of anomalies but from increasing complexity and technology-driven opportunity.

To balance the enthusiasm, the next few chapters look at what we *cannot* do with digital technology, at least not today. This requires explaining three classic concepts that emerged in the 20th century: Shannon's information theory, the Church-Turing thesis, and Gödel's incompleteness of formal models. In the later chapters, I consider the concept of determinism and examine how we can build models that embrace uncertainty using the notion of probability. Along the way, I need to confront another paradigm that emerged in 20th century called digital physics and a view that human cognition is software.

This part of the story begins in chapter 7, where I examine the concept of information — what it is and how to measure it. In this chapter, I introduce Claude Shannon's way of measuring information and show that his notion of information often cannot be represented digitally. I define an "information-processing machine" more broadly than what can be realized using software and computers, as they exist today.

In chapter 8, I explain what software cannot do. I point out that the number of information-processing functions is vastly larger than the number of possible computer programs. I introduce Alan Turing's undecidability result, which shows that useful information-processing functions exist that are not realizable by software on today's computers. But it does not follow that if a function is not realizable by software, then it is not realizable by any machine.

I caution against getting carried away by enthusiasm, marveling at what has already been accomplished with software, and caution against predicting that natural phenomena such as cognition and understanding are realizable in software. Here, I am forced to confront a belief that some people call "digital physics": that the physical world is somehow software or equivalent to software. I argue that this idea is unlikely to be either true or useful as a way of understanding the physical world, at least in its more extreme forms, and I show that this thesis is not falsifiable and therefore not scientific.

In chapter 9, I go beyond the countable world of computing and argue that computers are not universal machines and their real power comes from their partnership with humans. I explain the notion of a continuum, a concept that is out of reach for software and rejected by digital physics but seemingly essential for modeling the physical world. I examine the fundamental limitations of formal models that underlie the world of software, and I argue that the partnership and coevolution of humans and computers is much more powerful than either alone. In this chapter, I explain Kurt Gödel's famous incompleteness theorems, which impose fundamental limits on any modeling formalism that is capable of self-reference. We need to be humble, but we also need to recognize the as yet vast unexplored potential that still waits for us to catch up.

In chapter 10, I consider determinism, a property of software and many mathematical models of nature. I argue that determinism is a property of models not of the physical world. But it is an extremely valuable property, one that has historically delivered considerable payoffs in engineering and science. However, determinism also has its limits. Even deterministic models may not be usefully predictive because of chaos and complexity. Also, families of deterministic models that embrace both discrete and continuous behaviors are incomplete. There are unavoidable holes where determinism breaks down, and deterministic models have their limitations. In many cases, nondeterministic models are simpler and better reflect what we do not know. Nondeterministic models, used explicitly and judiciously, play an essential role in engineering.

In chapter 11, I finally confront the meaning of randomness and its measure, probability, which quantifies the likelihood of nondeterministic events. I argue that probability is fundamentally a model of *uncertainty* about something and not directly a model of that something. It models what we do not know. I examine the long-standing debate between the frequentists and the Bayesians, coming down solidly on the side of the Bayesians. I show that the philosophical difficulties presented by randomness vanish when using models in the engineering sense rather the scientific sense and when interpreting probability in the Bayesian sense. In this chapter, I also reconsider continuums and argue that probabilistic models over continuums reinforce the conclusion that digital physics is extremely unlikely. As a consequence, we should demand incontrovertible evidence for digital physics before accepting it.

In the final chapter, I tie things together by examining the epistemic role that models have in technology and the relationship between models and the physical systems they ultimately model. I leverage the previous arguments in the book: At least with digital technology, so many layers of abstraction exist between the models and the physical reality that the connection between the two becomes tenuous indeed. Moreover, the self-scaffolding that software paradigms have, described in chapter 5, allows these models to stand on their own, almost but not completely independent of physical reality. I argue that this does not lead to a Cartesian mind-body dualism, but it does emphasize the need to insist, with great determination and discipline, on separating the map from the territory. Models are best viewed as having a separate reality from the physical world, despite existing in the physical world.

The most expressive modeling paradigms are capable of self-reference, which enables them to build their own scaffolding but also makes them necessarily incomplete. This incompleteness is fundamentally what enables creativity and ensures that what we can accomplish with technology is limitless. So what holds us back? In this final chapter, I consider both the obstacles to progress and the threats that technology, when misapplied, can have on society.

Acknowledgments

The author gratefully acknowledges contributions and helpful suggestions from Christopher Brooks, Malik Ghallab, Thomas Henzinger, Madeline Johnson, Hokeun Kim, Gil Lederman, Marten Lohstroh, Dave Messerschmitt, Mehrdad Niknami, Rodion Rathbone, Rhonda Righter, Bernhard Rumpe, Naresh Shanbhag, Joseph Sifakis, Marjan Sirjani, Kimball Strong, David J. Stump, and Eli Yablonovitch. I would also like to thank three anonymous reviewers commissioned by the publisher who were extremely helpful. Several of these people disagreed with major points that I make in the book, and they thereby helped me to understand where my arguments needed to be strengthened or reworked. All remaining errors and opinions that I have stubbornly stuck to are entirely my own, not those of these contributors.

Most especially, however, I would like to thank two very special people who played a major role in the development of this book. The first is Heather Levien, who, unlike me, really knows how to write and without whom this book would be a disorganized pile of random ideas. The second is my mom, Kitty Fassett, a professional musician with an aversion for mathematics but a true intellectual and also a great writer. Without her help, this book would be unreadable to nonspecialists. She was my guinea pig, telling me each place where a nonspecialist might get lost.

I also thank the staff at MIT Press and Heather Jefferson for her superb copy editing. In addition, I thank the many unwitting contributors who have offered their thoughts through largely anonymous media such as Wikipedia and the contributors who have generously posted images online that I can (and have) reused because of their choice of creative commons licenses.

YANG

1 Shadows on the Wall

··· in which I examine the very idea of "facts" and "truths," showing that: collective wisdom about them can be better than individual wisdom; a narrative about facts can be more interesting than the facts themselves; facts and truths may be invented or even designed, not just discovered; facts and truths may be wrong; and it can cost billions to show that facts are true. And, oh yes, nerds are misunderstood, and science and engineering get confused.

1.1 Nerds

I am a nerd. According to the Merriam-Webster dictionary, a nerd is

> an unstylish, unattractive, or socially inept person; especially: one slavishly devoted to intellectual or academic pursuits.
> a person who is very interested in technical subjects, computers, etc.

Who but a nerd would start a book with a quote from the dictionary? I wouldn't expect a nerd to write very well, particularly not for a general audience. Actually, I'm quite sure this book would be much better if it were written by someone else. But I can't get anyone else to write it, so I will compensate by quoting the writing of others, even from dictionaries.

The previous definition was presumably written by a trustworthy expert on the subject of nerdiness. We expect that the publishers of dictionaries go to some effort to ensure that the definitions are written by experts. In contrast, we cannot assume that a Wikipedia page about nerds would be written by experts. Anyone with Internet access can modify the contents of a Wikipedia page. There is no vetting of expertise. Nevertheless, the page for "nerd" largely concurs with Merriam-Webster but offers more:

> Though originally derogatory, "Nerd" is a stereotypical term, but as with other pejoratives, it has been reclaimed and redefined by some as a term of pride and group identity.

Now I am reassured that I can be proud to be a nerd. Continuing with an etymology,

> The first documented appearance of the word "nerd" is as the name of a creature in Dr. Seuss's book *If I Ran the Zoo* (1950), in which the narrator Gerald McGrew claims that he would collect "a Nerkle, a Nerd, and a Seersucker too" for his imaginary zoo. [citations to Merriam-Webster and the American Heritage Dictionary] The slang meaning of the term dates to the next year, 1951, when *Newsweek* magazine

reported on its popular use as a synonym for "drip" or "square" in Detroit, Michigan. ... At some point, the word took on connotations of bookishness and social ineptitude.

I like this narrative better than the dictionary definition because it focuses on how the word came about and how it evolved rather than what it is. Most facts are more interesting when we understand how they came to be facts rather than just accepting them as if they were always there.

But is this Wikipedia article authoritative? The article points out that American satirist "Weird Al" Yankovic's song "White and Nerdy" states that editing Wikipedia is a stereotypical nerd interest. I am therefore reassured that this article is likely written by experts on nerdiness.

There is a big difference in style between a dictionary definition and a narrative about culture. A dictionary definition is usually understood to give a fact, a truth. Merriam-Webster defines "definition" as "an explanation of the meaning of a word, phrase, etc." This definition gives definitions the aura of facts and truths, deemphasizing their instability and fluidity with human culture.

Most of us approach technology as if it too were a compendium of facts and truths. We assume that technology advances because people discover more facts and truths. Because the discovery of facts and truths is the realm of science, science therefore drives technology. But I doubt that the technology of Wikipedia came about as a consequence of the discovery of facts and truths.

Wikipedia is a software system created by Jimmy Wales and Larry Sanger, who put the first version online in 2001. I don't know them, but one of my prejudices is that many software people are nerds, so there is a reasonable chance they too are nerds.

According to the Wikipedia article on Wikipedia, Sanger coined its name as "a portmanteau of wiki and encyclopedia." Following the link to the page on "wiki," we learn that a wiki is a website that "allows collaborative modification of its content and structure directly from the web browser." That page tells us that "wiki" is a Hawaiian word meaning "quick" and credits Ward Cunningham with inventing the wiki. I hope you will agree that it would be odd to say that Cunningham "discovered" the wiki, so presumably this was not an advance creditable to science. But the nuanced relationship between discovery and invention and between science and engineering is not always so clear.

Cunningham's 2001 book with Bo Leuf describes the wiki concept as follows:

> A wiki invites all users to edit any page or to create new pages within the wiki Web site, using only a plain-vanilla Web browser without any extra add-ons. Wiki promotes meaningful topic associations between different pages by making page link creation almost intuitively easy and showing whether an intended target page exists or not. A wiki is not a carefully

crafted site for casual visitors. Instead, it seeks to involve the visitor in an ongoing process of creation and collaboration that constantly changes the Web site landscape. (Leuf and Cunningham, 2001)

I love that web browsers come in "plain vanilla" flavors. I wonder what other flavors are available.

This description starts to give us the sense that a wiki, and particularly Wikipedia, is as much a cultural artifact as a technological one. And as with all cultural artifacts, it didn't suddenly pop into existence at the instant of invention. In fact, inventions almost never do and nearly always have a strongly cultural element. The "meaningful topic associations between different pages" were in fact already present in the World Wide Web, which according to its Wikipedia article was "invented by English scientist Tim Berners-Lee in 1989."

It is interesting that this article identifies Sir Timothy John Berners-Lee (he was knighted by Queen Elizabeth II for his work) as a "scientist." His recognized contributions were certainly not of the nature of discovery of facts and truths and certainly not about the natural world, the main focus of science. Berners-Lee did get a bachelor of arts degree in physics, unquestionably a science subject, from Oxford, so I suppose calling him a scientist is justified. But I see no evidence that he is a successful scientist.

Or maybe I am misunderstanding what it means to be a scientist. Returning to trusty old Merriam-Webster, a scientist is "a person who is trained in a science and whose job involves doing scientific research or solving scientific problems." Hmm... Not very helpful. Looking up "science," we find it is "knowledge about or study of the natural world based on facts learned through experiments and observation." By this definition, if Berners-Lee's career goal was to study the natural world, then I would have to conclude that his career has not (yet) been very successful. If, in contrast, his career goal was to invent and engineer artifacts that had never before existed, then he has been spectacularly successful, richly deserving the knighthood. He created mechanisms that are today used by nearly every person in the developed world. He *changed* the world.

Berners-Lee's contributions are arguably more cultural than technical. The cultural context of the web and Wikipedia goes back even further. Vannevar Bush,[1] in a 1945 article "As We May Think," states,

---

[1] Vannevar Bush, in the Wikipedia article on him, is identified as an "engineer, inventor and science administrator." Bush, who died in 1974, was a towering figure. He was an MIT professor, dean of the MIT School of Engineering, and founder of Raytheon, a major U.S. defense contractor. During World War II, Bush coordinated several thousand scientists in the application of science to warfare. He started the Manhattan Project, which led to the development of nuclear weapons. At the end of World War II, Bush pressed for increased government support for science. His arguments led to the creation of the National Science Foundation (NSF), which today is one of the premier supporters of research in science and engineering.

Wholly new forms of encyclopedias will appear, ready-made with a mesh of associative trails running through them, ready to be dropped into the memex and there amplified. (Bush, 1945)

The memex is Bush's hypothetical microfilm viewer that has a structure analogous to that of hypertext, the essential feature of Berners-Lee's web. Berners-Lee's technical contribution was to make Bush's vision a reality.

So why is Berners-Lee identified as a scientist? Possibly whoever wrote the Wikipedia article intended this as an honorific in the sense that "engineer" would not be. In his wonderful book, *The Black Swan*, from which I get the title of this book, Nassim Taleb used the term "Platonicity" for the "desire to cut reality into crisp shapes" (Taleb, 2010). My classification of people, including myself, as "engineers" or "scientists" (or even as "nerds") stems from such Platonicity. But humans are complex and defy classification. You may be surprised that I, a nerd, am also an amateur artist (see figure 1.1).

Taleb argues that Platonicity, the desire to categorize, the obsessive focus on taxonomy, "makes us think that we understand more than we actually do."

What I call Platonicity, after the ideas (and personality) of the philosopher Plato, is our tendency to mistake the map for the territory, to focus on pure and well-defined "forms," whether objects, like triangles, or social notions, like utopias (societies built according to some blueprint of what "makes sense"), even nationalities.

The arbitrariness of categories such as "scientist" and "engineer" is an example of Platonicity. It makes us sanguine in our understanding of the world, but it can be

Figure 1.1
Self-portrait of a nerd. Acrylic on canvas (2007).

misleading. In the rest of this chapter, I will focus on the difficulties in distinguishing discovery from invention, invention from design, and scientist from engineer.

1.2 Artificial and Natural

Herbert Simon, a hugely influential twentieth-century thinker and winner of both the Turing Award in computer science and the Nobel Prize in economics, in his book, *The Sciences of the Artificial*, makes a distinction between "artificial" and "natural" phenomena:

> The thesis is that certain phenomena are "artificial" in a very specific sense: they are as they are only because of a system's being molded, by goals or purposes, to the environment in which it lives. (Simon, 1996)

A system is artificial if it is "being molded, by goals or purposes." But in this statement, who does the molding? Simon presupposes it is humans. It could instead be God or some other teleological cause, and then no distinction would exist between the "sciences of the natural" and the "sciences of the artificial." One could even take as a definition of God as He who molds, by goals or purpose, our entire natural world. But there *is* a distinction between the artificial and the natural. A big one, Simon says.

Simon's examples of artificial phenomena include political systems, economies, engineered artifacts, and administrative organizations. The "molding" of such systems "by goals or purposes" is the process of *design*.

> Everyone designs who devises courses of action aimed at changing existing situations into preferred ones. (Simon, 1996)

Engineering, as a discipline, is fundamentally about design in this sense. Wikipedia, which as you may have realized by now, is my first recourse for many research problems, defines engineering this way:

> Engineering is the application of mathematics, empirical evidence and scientific, economic, social, and practical knowledge in order to invent, innovate, design, build, maintain, research, and improve structures, machines, tools, systems, components, materials, and processes. (retrieved March 1, 2016)

It then points out the obvious, "the discipline of engineering is extremely broad," and gives the origin of the term:

> The term *Engineering* is derived from the Latin *ingenium*, meaning "cleverness" and *ingeniare*, meaning "to contrive, devise."

Note that the word is *not* derived from "engine," as many people might assume. Instead, "engine" is derived from the same Latin roots.

Wikipedia effectively leverages recent triumphs of engineering. But is it authoritative? It creates *collective* wisdom, subjugating the role of *individual* experts, If you are old enough, you may remember the encyclopedias of the twentieth century, such as the *Encyclopedia Britannica*. According to Wikipedia, the *Britannica*

> is written by about 100 full-time editors and more than 4,000 contributors, who have included 110 Nobel Prize winners and five American presidents.

The Britannica is built in a very different way than Wikipedia. The editors recruit top experts to contribute to articles. The focus is on the *individual* experts, who, through their reputation, lend authority to the text.

Not Wikipedia. Anyone can edit a Wikipedia page. So how can these pages have any authority? At one level, Wikipedia replaces authority with accountability. Figure 1.2 shows the edit history of the Wikipedia page for Engineering quoted earlier. Notice near the bottom of the figure the most recent edit of this page, on

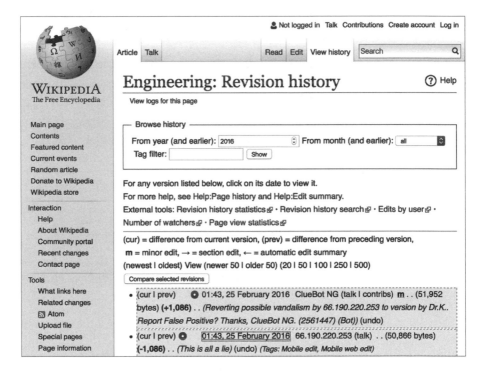

Figure 1.2

Edit history for the Wikipedia page on Engineering, retrieved March 1, 2016.

February 25, 2016. This edit is annotated with the comment, "Reverting possible vandalism." Indeed, the previous edit, which was made less than a minute earlier, has the comment, "This is all a lie." That edit removed quite a bit of text, including the previous definition of Engineering, and replaced it with, "This is al a lie" [sic]. The entire history of these edits is accessible on the Wikipedia site.

So who reversed the vandalism? The user is identified as "ClueBot NG." Clicking on that name reveals the page shown in figure 1.3. It turns out that ClueBot NG is a "bot," which Wikipedia defines as "a software application that runs automated tasks (scripts) over the Internet." On the page in figure 1.3, listed as item 8, is a description

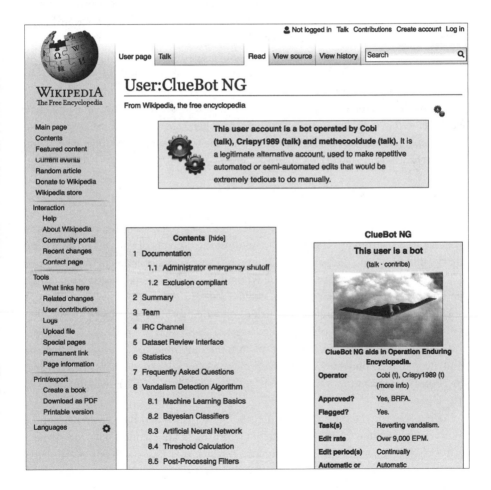

Figure 1.3
User page for the most recent editor shown in figure 1.2, retrieved March 1, 2016.

of the "Vandalism Detection Algorithm," which evidently is a piece of software that classifies an edit as either "vandalism" or "not vandalism," and if it is vandalism, it reverses the edit. The methods used to perform classification are statistical machine learning methods, linchpins of the currently hot area of data science. I will examine how this mechanism is analogous to an immune system and why that is important in chapter 9, and I will explain the principles behind how it works in chapter 11.

It is worth pausing and reflecting on how profoundly different all of this is compared with a twentieth-century encyclopedia. I will have more to say about this later, but one of the trends of the twenty-first century is the subjugation of the individual expert, the high authority, the intellectual hero. In the twentieth century, the phrase "triumphs of modern physics" would evoke in our minds Einstein, Bohr, Schrödinger, Heisenberg, and a few others. Of course, many others contributed, many of them also recognized as heroes. Experts do contribute to Wikipedia articles, but their text may be modified and elaborated on by anyone, including vandals. And readers rarely check to see who wrote the text. The authority of the author seems to be irrelevant. The text reflects a collective wisdom, not an individual one.

We now face an interesting conundrum. Which is closer to the truth, the collective wisdom or the individual one? This question gets ensnarled by what we mean by "truth." The answer is different if truths can be created rather than just being discovered.

1.3   Design and Discovery

In contrast to Simon's "sciences of the artificial," the "sciences of the natural" study what nature has given us. The goal is to uncover the "secrets of nature," presupposed to exist disembodied, independent of humans. These secrets occupy neither time nor space; they do not come into existence when they are discovered, occupying a time, nor do they exist at the place they are discovered.

The idea that these secrets exist disembodied dates back at least to Plato, who postulated ideal "Forms," objective and eternal truths that are impossible to know completely. Plato states that these Forms are the only objective truths, and that the ultimate goal of a "philosopher" (a lover of knowledge) is to understand these forms. The Forms represent the most accurate reality, and Plato calls knowledge of them "the Good."

Plato's Allegory of the Cave (see figure 1.4) suggests that human perception of reality is always imperfect. In the allegory, prisoners are chained with their backs to a low wall (bottom right in the figure). Their heads face the blank wall of the cave on which shadows are cast (top right) from a fire (top center). The shadows are of puppets and figures manipulated from behind the low wall, and the shadows on the cave wall constitute the only knowledge of reality that the prisoners experience. If a prisoner is released and can face the fire creating the light, then he will resist

Figure 1.4

Plato's Allegory of the Cave by Jan Saenredam, 1604. [ⓒThe Trustees of the British Museum.]

accepting the reality of the fire or the puppets casting shadows. If the prisoner is further allowed to exit the cave, then he will be blinded by the sun and further resist the reality of that external "ideal" world. And if a prisoner does accept some of the truths he is exposed to and tries to convey those truths to the prisoners behind the low wall, then they will reject his ideas as absurd.

The allegory underscores the difficulty of achieving the Good and explains why those individuals who do successfully convince others of some newly discovered, seemingly objective truth, the Einsteins and Bohrs of the world, are eventually deemed to be heroes.

Simon contrasts these objective truths with those of artificial phenomena. "If natural phenomena have an air of 'necessity' about them in their subservience to natural law, artificial phenomena have an air of 'contingency' in their malleability by environment." Their "subservience to natural law" presupposes a Platonic, disembodied existence of that natural law, regardless of whether it can be known.

A contrasting view is that laws of nature are models, created by humans, for how the physical world works. This view has gained considerable currency in the last few decades, perhaps beginning with Thomas Kuhn's groundbreaking and controversial

1962 book, *The Structure of Scientific Revolutions*, which postulated that scientific theories are framed by "paradigms," which are very much human ways of thinking about the world.

In either case, calling natural laws "laws" is a bit odd. It's almost as if nature is required to follow them, as citizens are required to follow the laws of a state. But what happens when nature violates a law of nature? Nature is not punished! Instead, the law becomes invalid. Imagine if a state worked that way. Each time a driver exceeded a speed limit, the speed limit would become invalid. But what about "laws of nature"? How can an ideal truth become invalid? And yet we've seen laws of nature become invalid many times.

David Deutsch, a British physicist at Oxford and a pioneer of quantum computing, in his 2011 book, *The Beginning of Infinity*, argues that science is more about "good explanations" than about laws of nature. Deutsch attributes these explanations to humans rather than some preexisting disembodied truth:

> Discovering a new explanation is inherently an act of creativity. (Deutsch, 2011, p. 7)

A "law of nature," in contrast, has a more humble role, that of codifying a connection between the past and the future:

> *[A]ny* purported law of nature — true or false — about the future and the past is a claim that they "resemble" each other by both conforming to that law. (Deutsch, 2011, p. 6, emphasis in the original)

Deutsch goes further to reject empiricism, arguing that neither laws of nature nor good explanations are derived from observation of the physical world:

> Experience is indeed essential to science, but its role is different from that supposed by empiricism. It is not the source from which theories are derived. Its main use is to choose between theories that have already been guessed. (Deutsch, 2011)

Here Deutsch echoes the thesis put forth earlier by the Austrian-British philosopher of science Karl Popper (1902–1994), who stated that these guesses, hypotheses about nature, arise from "creative intuition" and can only be tested empirically *after* they have been advanced (Popper, 1959). Deutsch points out that most of reality in the physical world is not directly observable to any human being. Black holes, quarks, and nuclear fusion in the sun involve scales, forces, and temperatures that no human has ever experienced. They are just as inaccessible as Platonic Forms, observable at best only as shadows on the wall. It cannot possibly be observations of black holes that lead to our theories about them, because we cannot observe them.

Plato recognized that the ideal truths of Forms could not be fully known by humans. But because they cannot be fully known by humans, isn't it more practical to view what we do know about nature as human-constructed models or what Deutsch calls good explanations? This would be more humble, tacitly acknowledging that even our most fervently held beliefs about nature are subject to improvement. I'm not saying that there are no truths, but just that we should always be required to question them.

Such humility is intrinsic in the Wikipedia model of collective wisdom. No individuals, no authorities, no matter how many accolades follow their names, can be trusted as the holders of "the truth." Knowledge should and must evolve. And as knowledge evolves, doesn't "the truth" also evolve? Of course, a realist might reply that it is *belief* not knowledge that evolves. For that realist, if a belief isn't true, it isn't (and never was) knowledge.

Deutsch points out that deference to authority was replaced during the enlightenment by empiricism, where ideas and theories are based on testing and experience. But he argues that experience alone is insufficient because so much of the physical world operates in conditions that cannot support human life and therefore cannot be directly experienced.

> Empiricism never did achieve its aim of liberating science from authority. It denied the legitimacy of traditional authorities, and that was salutary. But unfortunately it did this by setting up two other false authorities: sensory experience and whatever fictitious process of "derivation," such as induction, one imagines is used to extract theories from experience. (Deutsch, 2011)

Again echoing Popper, he argues that rather than deriving knowledge from experience, knowledge comes from a fundamentally creative process of conjecture, followed, often much later, by cleverly devised experiments that support the guesses (usually by failing to falsify them). Such clever experiments usually observe only indirect side effects. In other words, knowledge is engineered.

The next chapter will focus on the role that models play in both science and engineering, but for now suffice it to say that there is a tension here between *design* and *discovery*. Some artificial phenomena are not explicitly designed but rather emerge accidentally from human activity. The field of economics, for example, is full of the study of such emergent phenomena. The rules by which these artificial phenomena operate are arguably discovered rather than designed. And although many other artificial phenomena *are* designed, surely they too are subject to natural laws, which in the Platonic view at least must be discovered. But our knowledge of the natural laws is imperfect, so we must construct models of those laws using mathematics, for example. Are not these models designed?

Sir Isaac Newton's laws of motion are now known to be violated by nature, in that they fail at relativistic speeds and quantum scales. Those laws take the form of mathematical formulae, such as Newton's second law,[2]

$$F = ma, \qquad (4096)$$

which states that force equals mass times acceleration. This looks like an expression of a Platonic Form, but it is wrong! Despite being wrong, we don't hesitate to say that Newton "discovered" it, and people would look at us funny if we said that Newton "invented" it.

For some phenomena, we do not hesitate to use the word "invention" for their discovery. Consider the transistor, credited to John Bardeen, Walter Brattain, and William Shockley of Bell Labs (see figure 1.5).[3] They got the 1956 Nobel Prize in Physics "for their researches on semiconductors and their discovery of the transistor effect." They demonstrated the transistor effect with a device made of gold and germanium, although today transistors are realized mostly using silicon crystals with carefully introduced impurities called dopants. Note the careful phrasing in the Nobel Prize citation, "their discovery of the transistor effect." Nobel Prizes are not issued for inventions, only for discoveries. Yet most of us would say that the transistor was invented at Bell Labs in the 1950s.

But actually, a U.S. patent was awarded to Julius Lilienfeld in 1930 for the 1925 invention of a type of transistor now called a field-effect transistor (FET) (Lilienfeld, 1930). So if the transistor effect is a Platonic Form, then it was actually discovered earlier by Julius Lilienfeld. But patents are not issued for discoveries, only for inventions. According to the U.S. Patent and Trademark Office, you cannot patent "laws of nature, natural phenomena, and abstract ideas." A patent may be issued for a "novel use" of a law of nature, of course, because presumably everything is subject to the laws of nature. Such a novel use is an invention, not a discovery.

To be fair to the Nobel Prize committee, Lilienfeld's FET was actually significantly different from the one developed by Bardeen, Brattain, and Shockley. The

---

2 My former PhD thesis advisor, Dave Messerschmitt, once told me that when you publish a book, every equation you put in the book cuts your readership in half. I will call this principle "Messerschmitt's Law," although Dave tells me he did not discover this law. But I first heard it from him. Throwing caution to the wind, I am putting in an equation, but in an attempt to have some discipline, I will number each equation with an estimate of the remaining readership. Here, I've assumed optimistically a starting readership of 8,192, so the presence of this equation has cut it to 4,096. The next equation will be numbered 2,048. These are powers of 2 to make it easier to evenly divide by 2 each time and to underscore that I really am a nerd. If and when I get down to equation (1), I can write whatever I want because I will presumably have no more readers. As a side note, my PhD thesis had several dozen equations in it. It makes me wonder whether Dave ever read it.

3 Shockley moved in 1956 from New Jersey to Palo Alto, California, and started Shockley Semiconductor Laboratory in what would later become known as Silicon Valley. Eight of Shockley's employees left his company in 1957, the year I was born, to found Fairchild Semiconductor, the first successful high-tech company in Silicon Valley. Many other Silicon Valley giants, including Intel, were founded by former Fairchild employees. Arguably, Shockley's move to Palo Alto was the founding of Silicon Valley.

Figure 1.5
John Bardeen, Walter Brattain, and William Shockley, who received the 1956 Nobel Prize in Physics for the discovery of the transistor effect.

Bell Labs transistor was of a type known today as a bipolar transistor. Interestingly, today, bipolar transistors are used only in niche applications because of their significantly higher energy requirements. FETs are used much more commonly, having become the workhorses for digital technology. An iPhone today contains billions of FETs.

It is not uncommon for inventions and discoveries to be messy in this way. The intellectual history of an idea is rarely clear, and yet, as a culture, we insist on singling out the "heroes" who bring these ideas to the fore.[4]

The tension between discovery and invention is not new. Is the transistor effect a "natural phenomenon" that has always existed, waiting to be discovered? As far as I know, nobody has ever found in nature sandwiches of gold and germanium or doped silicon operating as transistors. So transistors must not be natural phenomena. But when these sandwiches are constructed by man, nature takes over and regulates the movement of electrons in the material so as to realize the transistor effect. So a transistor appears to be a novel use of a law of nature.

But the movement of electrons in crystalline materials with impurities was not well understood until the transistor had been fabricated and studied; in fact, the phenomenon continues to be better understood to this day under more study. Is a natural law that had no manifestation in nature prior to a human-constructed invention, and that was not understood as a natural law until after such construction,

---

4 And then become embarrassed when those heroes disappoint us, as Shockley did by becoming a proponent of eugenics later in his life.

a Platonic Form? It seems to me questionable to claim that the transistor effect exists and has always existed, timeless and disembodied. To me, this stretches my understanding of the word "exists" beyond the breaking point.

This debate about natural laws dates back a long time. Aristotle, a student of Plato's, questioned the Platonic ideal Forms, arguing that knowledge is based on the study of *particular* things, and that generalizations arise from that study rather than preexisting in a disembodied Form. Aristotle used the term "natural philosophy" for the study of phenomena in the natural world, what we now call "science." Aristotle's world of facts is extensible; it can grow with study of the natural world. Plato's world of facts is fixed; all the facts are there as Forms, with many of them still waiting to be discovered.

But it seems that facts can become wrong. Despite being wrong, Newton's second law is amazingly useful. Engineers use it all the time to design cars, airplanes, robots, bridges, toys, and so on. It provides a good model for how *particular* things behave, as long as they are not traveling near the speed of light and are not being examined at subatomic scales. It cannot be a Platonic Form because it is violated by nature. It can, however, constitute knowledge in the Aristotelian sense because it generalizes nicely the behavior of macroscopic objects.

1.4 Engineering and Science

Simon says that design is about "changing existing situations into preferred ones." But what do we mean by "preferred" situations? In political systems, this may be highly subjective. In engineered systems, it may be much more objective. A political leader may prefer a situation where all immigrants are kept out, even when there is no objective evidence that this makes anything better for anyone. Engineers, by contrast, are often called on to defend their preferences with objective measures, such as lower cost or reduced energy consumption. Simon's "preferred situations" are open. But it is not uncommon in popular culture to assume that engineers primarily optimize preexisting designs. A somewhat silly joke underscores this point:

> **Question**: What is the difference between an optimist, a pessimist, and an engineer?

> **Answer**: An optimist sees a glass half full. A pessimist sees a glass half empty. An engineer sees a glass that is twice as big as it needs to be.

This joke plays on our preconception that engineers prefer whatever costs less. Many engineered systems, however, are "preferred" despite lacking any objective measures showing them to be better than preceding "existing situations." Apple's iPhone, for example, did not make phone calls better than its Nokia predecessors, and its battery life was distinctly shorter. And it certainly didn't cost less! It was "preferred" because of nonobjective properties. It was fundamentally a *creative* contribution to humanity, not an optimization. And yet it was most certainly an engineered artifact.

When using the phrase "sciences of the artificial" for the creation and study of human-made phenomena, Simon laments the pejorative connotations of the term "artificial," saying "our language seems to reflect man's deep distrust of his own products." Arguably, distrust of the products of nature is equally justified, as suggested by the poem on page 17. But as a father of teenagers with iPhones, I can attest that distrust of the artificial is real. Trusted or not, there is no question that smartphones are transformative. The iPhone (and its subsequent competitors), together with other recent innovations in wireless communications and computer systems, enable us to carry nearly all of human published information in our pockets. To call this "transformative" seems like an extraordinary understatement. It is much more a triumph of engineering, "the sciences of the artificial," than of the sciences of the natural. Yet there is little, if any, invention in the iPhone. Nearly every important aspect of the phone existed already in other products when it was introduced. The iPhone is much more the result of design than either invention or discovery.

UNNATURAL

I'm sure Nature has disapproved of me
for years, as if it had overheard
one of my silent screeds against it,
and my insistence that only the artificial
has a real shot at becoming more
than we started with, designed,
revised, something completely itself.
If it could speak, Nature might say
it contains lilies, the strange beauty
of swamps, the architectureal art
of spiders, the many et ceteras
that make the world the world.
Nothing man-made can compete,
Nature might say. Oh Nature
has been known to go on and on.
And if it wanted to push things further,
it could cite our sleek perfection

of bombs and instruments of torture,
our nature so human we hide
behind words that disguise and justify.
But that's as generous as I want to be
in giving Nature its say. I've seen it
randomly play its violence card —
natural, no-motive crimes
with hail and rain and vicious winds,
taking out, say, trailer courts and
playing fields and homes for the elderly.
So I want to be heard and overheard,
this time for real, out loud, in fact
right in Nature's face, to say I prefer
the artifice in what's called artificial,
the often concealed skill involved,
without which we'd have no accurate
view of ourselves, or of lilies in the pond.

— Stephen Dunn

Yet in current Western culture, it seems that most people respect an inventor more than an engineer and a scientist more than an inventor. Colin Macilwain, in an article in *Nature*, attributes to William Wulf, former president of the U.S. National Academy of Engineering, the following statement:

> There is a general attitude among the scientific community that science is superior to engineering. (Macilwain, 2010)

This attitude spills out from the scientific community to the general culture. We use the term "rocket scientist" for extremely smart people, although most of what the people who put together the space program do is engineering. Macilwain goes on,

> Wulf attributes this partly to the "linear" model of innovation, which holds that scientific discovery leads to technology, which in turn leads to human betterment. This model is as firmly entrenched in policy-makers' minds as it is intellectually discredited. As any engineer will tell you, innovations, such as aviation and the steam engine, commonly precede scientific understanding of how things work.

It is hard to point to any scientific discovery that led to the iPhone, in the sense that every scientific discovery it depends on was already in widespread use in other products. Nevertheless, it is easy to find evidence that popular culture assumes that this linear model of innovation is in fact how things work. For example, About.com, an advertising-funded website centered around articles on a huge variety of subjects, collects reader commentary. On the question of "Engineer vs Scientist - What's the Difference?" some of the reader answers are:[5]

> Scientists are the ones who create the theories, engineers are the ones who implement them. They compliment [sic] each other...

> Science is a lot of high level theory and engineering is implementation and optimization.

> Engineers deal with math, efficiency and optimization while Scientist [sic] deal with "what is possible."

> Engineers trained [sic] for Using tools, where Scientists are trained for Making them.

> Scientists develop theories and work to verify them, Engineers search in these theories to "optimize" things in real life.

---

5 http://chemistry.about.com/od/educationemployment/fl/Engineer-vs-Scientist-Whats-the-Difference.htm, updated June 29, 2015, retrieved March 1, 2016.

A scientist invents a law and an engineer applies it.

Scientist for invention of new theories [sic]. Engineers for applying those theories for piratical [sic] applications.

These views are clearly not authoritative, but rather are reflective of popular perception. Note the contrast between the style of this website and that of Wikipedia. This one is a portal for *individual* wisdom (and stupidity), whereas Wikipedia is a portal for *collective* wisdom.

Kuhn, a highly regarded historian of science and a philosopher, in his 1962 book, *The Structure of Scientific Revolutions*, echoed what Wulf claimed was the "general attitude among the scientific community," stating that certain kinds of scientific measurement tasks are "hack work to be relegated to engineers or technicians" (Kuhn, 1962). To Kuhn, clearly engineers were a rung down the ladder from scientists. But like his repeated reference in the same text to scientists as "men," we can forgive this disparaging remark about engineers because at the time he was writing, this view was standard in contemporary culture and had a strong element of truth.

Kuhn addressed the question of what is science, stating, "to a very great extent the term 'science' is reserved for fields that do progress in obvious ways." But, he points out, many fields progress in obvious ways:

Part of our difficulty in seeing the profound differences between science and technology must relate to the fact that progress is an obvious attribute of both fields.

Kuhn rejects the pervasive idea that the progress of science is toward some Platonic truth:

We may ... have to relinquish the notion, explicit or implicit, that changes of paradigm carry scientists and those who learn from them closer and closer to the truth.

If truth is not the goal, then what gives "progress" its directionality? Kuhn postulates that science may actually have no goal, an observation that he recognizes will be difficult for many people to swallow.

We are all deeply accustomed to seeing science as the one enterprise that draws constantly nearer to some goal set by nature in advance.

He then draws an analogy between the progress of science and Darwin's theory of evolution:

The *Origin of Species* recognized no goal set either by God or nature.

The lack of a goal for science may be a shock, but for technology, it seems easier to accept. It is hard to postulate any ultimate Platonic "truth" of technology, any goal that when reached finishes the field. Technology progresses if once it is known how to make certain things, this knowledge is not forgotten.

In a 1984 book, the philosopher John Searle supports Wulf and Kuhn about this twentieth-century view of science:

> "Science" has become something of an honorific term, and all sorts of disciplines that are quite unlike physics and chemistry are eager to call themselves "sciences." A good rule of thumb to keep in mind is that anything that calls itself "science" probably isn't — for example, Christian science, or military science, and possibly even cognitive science or social science. (Searle, 1984, p. 11)

Spencer Klaw, in his 1968 book, *The New Brahmins — Scientific Life in America*, writes about "the awe that scientists now inspire," where

> science has become a form of established religion, and scientists its priests and ministers. (Klaw, 1968, p. 12)

Many disciplines seek to emulate the methods of science, hoping for similar payoffs. The "scientific method," where a hypothesis is formed and experiments are designed to attempt to falsify the hypothesis, is useful in many disciplines that have little connection with natural science. But the value of the scientific method is often not as great in these nonsciences. Referring to social science, Searle observes, "the methods of the natural sciences have not given the kind of payoff in the study of human behavior that they have in physics and chemistry" (Searle, 1984, p. 71).

Popper, before Kuhn, stressed that the core of the scientific method is falsifiability. A theory or postulate is scientific only if it is falsifiable, according to Popper. To be falsifiable, at least the possibility of an empirical experiment that could disprove the theory must exist. For example, the postulate that "all swans are white" is not supported by any number of observations of white swans. But the postulate is falsifiable because an experiment *may* find a black swan. Hence, it is a scientific theory, albeit a false one.

Kuhn rejects Popper's conclusion that a scientific theory is rejected by falsification, arguing that even in the face of evidence against it, a theory will not be rejected until a replacement theory is invented:

> [T]he act of judgment that leads scientists to reject a previously accepted theory is always based upon more than a comparison of that theory with the world. The decision to reject one paradigm is always simultaneously the decision to accept another, and the judgment leading to that decision involves the comparison of both paradigms with nature and with each other. (Kuhn, 1962, pp. 77–88)

Kuhn is saying that even if an experiment seems to falsify a hypothesis, scientists will not reject the hypothesis until they have a replacement hypothesis. He says, "If any and every failure to fit were ground for theory rejection, all theories ought to be rejected at all times."

Popper's emphasis on using experiments to falsify hypotheses is healthy. Well-constructed experiments undermine astrology, phrenology, and many other pseudo-sciences. But as Kuhn points out, experimental evidence is always subject to interpretation. If there is no new paradigm aligning with the experiments, then the experimental results are more likely to be viewed as errors than falsifications.

Experiments are also useful in the "sciences of the artificial." Engineers and computer scientists do perform experiments but not usually with an eye toward falsification or to compare against nature. The mere fact that you do experiments does not make you a scientist.

In its narrower usage, as reflected by the Merriam-Webster definition previously quoted, the word "science" refers to the study of nature, not to the study or creation of artificial artifacts. Under this interpretation, many of the disciplines that call themselves a "science," including computer science, are not, even if they do experiments and use the scientific method.

To be sure, beginning with the information technology revolution in the 1990s, the role of engineering has been changing. I believe that this is because digital technology and software have created an explosion of possibilities in the "sciences of the artificial." There is nothing natural about being able to communicate instantaneously with another person nearly anywhere on the planet. There is nothing natural about being able to see inside the human body. There is nothing natural about being able to carry all of human published information in your pocket. These are all the results of engineering more than science. More important, they are *creative* products, not inevitable consequences of scientific discovery.

Nevertheless, science still captures our imaginations and delivers spectacular results. The announcement on February 11, 2016, of the detection of gravitational waves emitted by colliding black holes, for example, got a great deal of press (see, e.g., the *New York Times* article by Overbye [2016]). Gravitational waves were predicted by Einstein more than a century ago, but detecting them has turned out to be astonishingly difficult. The announced detection was accomplished by the Laser Interferometer Gravitational-Wave Observatory (LIGO), at a cost of approximately $1.1 billion. The detected wave lasted one fifth of a second, and analysis indicates that it was produced by a collision between two black holes a billion light years away. This style of science is unlikely to have the practical consequences that early twentieth-century science had. It is "pure science," in that it seeks knowledge for its own sake.

As might be expected, the high cost of this project has drawn some criticism. Horgan (2016) subtitled his column that reported this result in *Scientific American*

> *Was the gravitational-wave experiment worth its $1.1-billion cost if it merely confirms that Einstein was right?*

In his article, he quotes chemist Ashutosh Jogalekar, who blogs as Curious Wavefunction:

> Some sources are already calling the putative finding one of the most important discoveries in physics of the last few decades. Let me not mince words here: if that is indeed the case, then physics is in bad shape.

Horgan goes on:

> In an email to me, a historian of technology was more blunt: "So a 100 year old theory has been confirmed experimentally—big whup. Did anyone think Einstein was wrong? There wasn't any controversy, was there? Was anyone credible claiming that spacetime isn't curved, or that black holes don't exist? I can get that this was quite an experimental trick and technological feat. But this isn't doing anything to convince me that public funds spent on this stuff wouldn't be better spent on medical research. Or clean fuels, or any number of things that would apply scientific expertise toward justice or the alleviation of human suffering.

The acknowledgment that this experiment was "quite an experimental trick and technological feat" is interesting. It raises the question, is the contribution of LIGO science or engineering? The basic method used, laser interferometry, has been understood by scientists as a way to measure gravitational waves since the 1970s. But building a system with adequate sensitivity was not easy.

Given that Einstein's model predicted gravitational waves 100 years ago, that there seems to be no controversy among scientists about the correctness of this prediction, and that the laser interferometry technique for measuring gravitational waves has been known for decades, it may appear that *no* new science resulted from the $1.1 billion investment. But it is probably not the validation of the existence of gravitational waves that is really the scientific contribution, but rather the demonstration of a new modality for observing events in the universe that were previously invisible to us. Specifically, this experiment has given the first observation of two black holes merging. That such events occur is perhaps not surprising, but most certainly intellectual value can be found in the first demonstration of a new kind of telescope into the universe that is capable of observing such events.

So instead, we should probably view the $1.1 billion as an investment in the *engineering* of a new device that can now enable a new form of astronomical observation. And the device is quite a triumph of engineering.

Let me try to explain the magnitude of the engineering challenge that the LIGO team faced. First, two detectors were built 3,000 kilometers apart so that the difference in time of arrival of gravitational waves at the two detectors would provide an indication of the direction of the source, and so that entirely independent observations could corroborate each other. Building two detectors can't be more than twice as hard as building one, so this was not the biggest challenge.

Each detector consists of an L-shaped ultra-high-vacuum cavity 4 kilometers long on each side (this alone is not easy to build; see figure 1.6). It uses laser interferometry to measure extremely slight distortions in space-time caused by passing gravitational waves; these distortions change the distance between the two ends of the 4-km cavity ever so slightly, by much less than the diameter of a proton! By measuring this change in distance, once all other possible causes for the change in distance have been eliminated, one can infer that the change in distance was caused by a passing gravitational wave distorting space-time. To minimize spurious causes for changes in distance, each detector has to be completely isolated from sources of vibration such as seismic events and human activity such as automotive traffic. Even the most minor such vibration would render the instrument useless.

It is hard to make the case that a gravitational wave telescope will improve (or even affect) the human condition in any tangible way. Nevertheless, the project may in fact have practical and tangible impact by contributing improvements in engineering methods. The ability to detect such extremely small variations in distance surely has applications elsewhere.

Figure 1.6
LIGO gravitational wave detector in Livingston, Louisiana. [Courtesy Caltech/MIT/LIGO Laboratory.]

NASA, whose main mission (I believe) is space exploration in the name of science, frequently uses their contributions to technology development as further justification for the expenditure on space exploration. They claim contributions to light-emitting diodes (LEDs), infrared ear thermometers, artificial limbs, ventricular assist devices, anti-icing systems for aircraft, safety grooving on highways, improved automotive tires, chemical detectors, land mine removal, firefighter gear, and many other technologies (NASA, 2016). To me, this reads as a substantial contribution to technology, irrespective of the contribution to science.

Assuming that LIGO is a triumph, who is the hero? The article announcing the measurement of gravitational waves in *Physical Review Letters*, published on February 11, 2016, has 1,019 authors (Abbott et al., 2016). The author list occupies 5 of the 16 pages of the article. It is hard to identify an "Einstein" from this list. According to the *Boston Globe*, Rainer Weiss, now a Professor Emeritus at MIT, is credited by many scientists with being the mastermind of the project, over significant protests from Weiss, who demurs that many people contributed a great deal (Moskowitz, 2016). Assuming Weiss is right, the LIGO project is a form of collective rather than individual wisdom, much like a Wikipedia article. And it is likely that most of these authors would self-identify as "scientists" and not as "engineers." To me, most if not all of these 1,019 authors are engineers as well as scientists, defying Platonicity.

An engineered artifact such as an iPhone is similarly a form of collective wisdom. It is impossible to identify all the individuals who contributed significant technical content to the iPhone, but I'm sure it is many more than 1,019.

In a famous essay, Leonard Edward Read (1898–1983), libertarian and founder of the Foundation for Economic Education (FEE), accounted for the technical contributions required to make a humble wooden pencil (Read, 1958). Written from the point of view of the pencil, it starts with, "Not a single person on the face of this earth knows how to make me." He then chronicles the processes and materials that go into fabricating a pencil:

> My family tree begins with what in fact is a tree, a cedar of straight grain that grows in Northern California and Oregon. Now contemplate all the saws and trucks and rope and the countless other gear used in harvesting and carting the cedar logs to the railroad siding. Think of all the persons and the numberless skills that went into their fabrication: the mining of ore, the making of steel and its refinement into saws, axes, motors; the growing of hemp and bringing it through all the stages to heavy and strong rope...

He goes on to explain how the wood is milled, kiln dried, and tinted; how the graphite is mined and then mixed with clay and sulfonated tallow; how the lacquer paint is made from castor beans and castor oil; how the label is made with carbon black mixed with resins; how the metal is mined and refined; and how the eraser is made from rape seed oil, sulfur chloride, rubber, pumice, and cadmium sulfide.

And an iPhone is much more complicated than a pencil. Evidently, even Steve Jobs wouldn't know how to make an iPhone (or even a pencil). In a reference to the "invisible hand" of the economist Adam Smith (1723–1790), Read continues:

> There is a fact still more astounding: The absence of a master mind, of anyone dictating or forcibly directing these countless actions which bring me into being. No trace of such a person can be found. Instead, we find the invisible hand at work.

Such an engineered artifact is an embodiment of collective wisdom even more extreme than Wikipedia, where at least a log is kept of the individual contributions.

Although we can't even trace the forces behind the invisible hand, widespread recognition exists that many of these forces are driven by people's technical skills. Cultivating such talent is a prerequisite for a modern economy. Today, policymakers and much of the public recognize the value in Science, Technology, Engineering, and Mathematics (STEM) education. This term bundles together a broad set of technical disciplines. It still puts Science first, but this may be as much about being able to pronounce the acronym as it is about relative priorities. Indeed, Liana Heitin blogs that STEM was originally SMET, which perhaps better reflected perceived priorities but was not so euphonious (Heitin, 2015).

Much of the political motivation in STEM may be pragmatic; it's more about being able to get jobs than it is about intellectual search. But we may be underestimating the intellectual search that is intrinsic in the "sciences of the artificial." Without the *engineering* tour-de-force of the LIGO project, we would not have humankind's first gravitational telescope. Maybe this telescope will reveal other colliding black holes and other phenomena that may help us better understand the origin of the universe. So indeed, sometimes engineering does precede science rather than the other way around.

We see many other indicators of a shifting attitude toward technology and engineering. In the twentieth century, an "institute of technology" would be viewed as primarily a vocational school rather than a center of intellectual activity. MIT and Caltech changed that notion, and we are even starting to see "technical high schools" emerge as much more than vocational training.

Technology and engineering are distinctly not about discovering preexisting, disembodied truths. They are about creating things, processes, and ideas that never before existed. Pursuit of the Platonic Good, the preexisting, fixed world of Forms, is no longer what is driving humanity forward. We are instead creating knowledge and facts that never before existed, embodied or not.

In the next chapter, I focus on the relationship between discovery and invention. A key theme of that chapter is to understand the role of models in engineering and science. My essential claim is that models are invented, and when those models are modeling physical phenomena, the corresponding physical phenomena, not the models, are discovered. And even those physical phenomena may be brand new, as was the case with the transistor.

2   Inventing Laws of Nature

··· in which I argue that models are invented, not discovered; that engineers and scientists use models in complementary, almost opposite ways; that all models are wrong, but some are useful; and that the use of models can slow as well as advance technological progress by establishing a backdrop of unknown knowns, by forcing increased specialization, and by requiring humans to assimilate new paradigms.

2.1   The Unknown Knowns

Drawn by its provocative title, I recently read *Inventing Nature*, a wonderful book by Andrea Wulf (Wulf, 2015). Wulf's book tells the story of Alexander von Humboldt (1769–1859), a remarkable Prussian about whom I had previously known nothing except that Humboldt county and Humboldt Redwoods State Park in California, along with numerous other places and things, had been named after him. Wulf boldly states, "Humboldt gave us our concept of nature itself." Easing somewhat my embarrassment at my ignorance, she goes on, "The irony is that Humboldt's views have become so self-evident that we have largely forgotten the man behind them." Further easing my embarrassment, Sandra Nichols (2006), in an article titled "Why was Humboldt forgotten in the United States?" reassures me that I am not alone. Nichols postulates many reasons for our collective amnesia, but to me the most poignant is "shifts in scholarship," where "the search for a comprehensive view of science was soon set aside in favor of specialization." In Germany, Humboldt and his brother Wilhelm have most definitely not been forgotten. The Humboldt University of Berlin is named after the two of them, and the brother, Wilhelm, is credited with establishing the "Humboldtian model of higher education," which integrates teaching in the arts and sciences with research. This is a defining principle of all top universities today.

In her claim that Alexander von Humboldt "invented nature," Wulf shows us that scientific truths can come into existence and then become part of the human psyche, background knowledge that we accept with such tenacity that we no longer think of them as scientific truths. They just are. Wulf summarizes Humboldt's breakthrough:

> Humboldt revolutionized the way we see the natural world. He found connections everywhere. Nothing, not even the tiniest organism, was looked at on its own. "In this great chain of causes and effects," Humboldt said, "no single fact can be considered in isolation."

Wulf credits Humboldt with being the first scientist to show evidence of human-induced climate change, for founding the field of ecology, and for articulating the first modern notion of "nature" itself. The most astonishing part of this story is that it never occurred to me that the connectedness of nature was not just a simple self-evident truth. It is so widely accepted today that it fades into the background of our basic instinct, along with notions of time and causality. Wulf explains the prevailing scientific thought at Humboldt's time:

> Inventions such as telescopes and microscopes revealed new worlds and with them a belief that the laws of nature could be discovered.

But Wulf points out that these laws of nature were understood one phenomenon at a time, as in Newton's laws of motion governing a falling object. Connectedness fell victim to reductionism.

Connectedness faded from our conscious approach to science into the unconscious, part of an unseen background, an unknown known. US Secretary of Defense Donald Rumsfeld, at a Department of Defense news briefing in 2002, made the following often quoted statement:

> ··· as we know, there are known knowns; there are things we know we know. We also know there are known unknowns; that is to say we know there are some things we do not know. But there are also unknown unknowns—the ones we don't know we don't know. And if one looks throughout the history of our country and other free countries, it is the latter category that tend to be the difficult ones [sic]. (Rumsfeld, 2002)

The Slovenian philosopher Slavoj Žižek pointed out that Rumsfeld didn't mention an obvious fourth category of knowledge, the "unknown knowns." These are the things we know but don't know that we know, which Žižek says is "precisely the Freudian unconscious" (Žižek, 2004). The interconnectedness of nature was, until I read Wulf's book, one of my unknown knowns. I didn't know that I knew that. Now I do, and thanks to Wulf, I also now realize that this "truth" that I know was not always known. She credits Humboldt with making it known.

Our unknown knowns bias our thinking. Thomas Kuhn, in his 1964 book *The Structure of Scientific Revolutions*, disrupted the prevailing view of science as "development-by-accumulation" (Kuhn, 1962). Instead of an accretion of discovered facts about the world, a scientific discipline is founded on a "paradigm," a conceptual framework that practitioners use, often unknowingly, to interpret observations and develop theories. Kuhn argued that these paradigms make scientific understanding necessarily subjective.

> Observation and experience can and must drastically restrict the range of admissible scientific belief, else there would be no science. But they cannot alone determine a particular body of such belief. An *apparently arbitrary element, compounded of personal and historical accident*, is always a formative ingredient of the beliefs espoused by a given scientific community at a given time. [emphasis added]

Kuhn's "arbitrary element" often takes the form of unknown knowns. A paradigm becomes so widely accepted and strongly held that its subjects no longer know it is there. They accept the paradigm as truth. As argued by Kant, the order we perceive in the world is shaped by our mind, which provides the distorting lens through which we perceive it. We impose order on nature rather than the other way around.

Kuhn's central claim is that scientific revolution, the truly momentous advances that occur from time to time, come about through paradigm shifts rather than through accretion of knowledge. The notion of scientific truth is therefore subjective, defined more by consensus of a scientific community than by Plato's ideal disembodied truth. An obvious corollary is that a paradigm is invented more than discovered.

Kuhn's position was highly controversial. It went very much against the grain of the prevailing philosophy of science, best articulated by Popper, which was about seeking objective truths. In 1965, a colloquium convened in London drew together many of the most prominent thinkers on the philosophy of science to respond to Kuhn's thesis. Imre Lakatos, a Hungarian philosopher of mathematics and science, and co-editor of the proceedings from the colloquium, wrote, "*in Kuhn's view scientific revolution is irrational, a matter for mob psychology*" (Lakatos, 1970, emphasis in the original). He goes on to criticize the notion of a "paradigm shift," writing that it is

> a mystical conversion which is not and cannot be governed by rules of reason and which falls totally within the realm of the *(social) psychology of discovery.* Scientific change is a kind of religious change. [emphasis in the original]

Despite these objections, Kuhn's notion of governing paradigms is useful for understanding the evolution of scientific thought. It is even more useful for understanding technology, where paradigms can be more obviously subjective.

In our modern technological world, our lives are governed by paradigms that are more obviously invented rather than discovered. These paradigms shape our understanding of the world, becoming unknown knowns. Consider, for example, the fact that in music and performing arts, it used to be that performer and observer had to be in the same room at the same time. Today, we can use Spotify and Hulu, among others, to carry much of the world's music and theater in our pockets, to be enjoyed whenever and wherever we choose. Most of us alive today have been able to listen to music without being in the same room with the musicians. This fact has changed the very meaning of the word "music" in ways that we don't notice. Think, for example, about the meaning of the phrase "my music." What would that have meant to a citizen of the nineteenth century, before Edison?

I was once telling my wife that a colleague at Berkeley, Miki Lustig, was leading a charge to teach many students, staff, and faculty at Berkeley about amateur radio, and to help them prepare to take the test to become licensed to operate amateur

Figure 2.1
Dilbert, an iconic nerd, in a cartoon by Scott Adams. [DILBERT ©1995 Scott Adams. Used by permission of UNIVERSAL UCLICK. All rights reserved.]

radios. My wife asked me why one would want to operate an amateur radio. The question had never occurred to me, and off the cuff, the best answer I could come up with was, "So they can communicate with anyone around the world." She asked me, "Why don't they just send email?" I saw a collision of paradigms, less momentous than the collision of two black holes, but nevertheless notable. Being able to communicate instantaneously with anyone around the world has become a background fact, an unknown known, a part of the technological paradigm through which we understand and manage our daily lives.

Paradigms change. Kuhn's scientific paradigms change relatively infrequently, and the changes can be quite disruptive to a scientific community. In *The Structure of Scientific Revolutions*, Kuhn quotes Max Planck:

> A new scientific truth does not triumph by convincing its opponents and making them see the light, but rather because its opponents eventually die, and a new generation grows up that is familiar with it.

The same thing happens with technological paradigms. Witness how much of modern technology becomes inaccessible to an aging brain. Our kids accept technological truths that are incomprehensible to older people. In fact, I believe that the pace of technological progress today is more limited by the inability of humans to absorb new paradigms than it is by any physical limitations of the technology.

Kuhn is generous to scientists whose paradigms are later supplanted by better paradigms:

> ⋯ those once current views of nature were, as a whole, neither less scientific nor more the product of human idiosyncrasy than those current today.

... If these out-of-date beliefs are to be called myths, then myths can be produced by the same sorts of methods and held for the same sorts of reasons that now lead to scientific knowledge.

We should be similarly generous to humans whose technological paradigms become obsolete, rather than thinking of them as luddites or dinosaurs. I would say to my kids, "Don't worry, you too someday will be a dinosaur."

Despite similarities, technological paradigms differ from scientific ones in significant ways. Technological paradigms are today much more diverse than scientific paradigms, reflecting immaturity of the field and rapid change. Kuhn argues that scientific paradigms are incommensurable. One paradigm cannot be understood or judged through the conceptual framework and terminology of the other. I will show in chapter 3 that this is less the case for technological paradigms, which may be layered in such a way as to interoperate. Nevertheless, incommensurable paradigms do arise, and it becomes necessary to build a metaparadigm within which to compare technological paradigms. I will attempt to do that through the notion of modeling.

Science, technology, and engineering are all built on models. Models are artifacts in the conceptual framework of a paradigm. Newton's second law, for example, is a model of the motion of an object subjected to a force. It takes the form of an equation, specifically equation (4096) on page 14, which has meaning in the paradigm of Newton's and Leibniz's calculus, the concept of force, and the Newtonian notion of time and space. If you studied physics in high school, you probably got brainwashed sufficiently that the concepts of force, time, and space are among your unknown knowns. But objectively, Newton gave no physical explanation for these concepts. Instead, he built a self-consistent and self-referential model where each of these concepts is defined in terms of the others, if defined at all.

Every engineered design is similarly a model, which can be as simple as a prototype of a physical shape or as complex as a million lines of code. Each such model has a meaning, a *semantics*, only within some modeling paradigm. And the modeling paradigm is all too often an unknown known, never articulated or consciously chosen. I will attempt now to break the logjam that is created by failing to recognize these unknown knowns.

2.2 Models of Nature

Merriam-Webster's online dictionary has no fewer than 14 definitions of the word "model." Only a few of these are relevant to how models are used in science and engineering:

4. a usually miniature representation of something; also: a pattern of something to be made

5. an example for imitation or emulation

   . . .

7. archetype

   . . .

11. a description or analogy used to help visualize something (as an atom) that cannot be directly observed

12. a system of postulates, data, and inferences presented as a mathematical description of an entity or state of affairs; also: a computer simulation based on such a system.

The first of these definitions is a concrete model, a material object in the physical world, whereas the last two are abstract models, where any material realization, for example, as ink on paper, is incidental. The two in the middle could be either concrete or abstract. Both abstract and concrete models help humans grasp concepts. Both kinds of models are created by humans. Models, therefore, can serve as a way for humans to record and communicate concepts.

For Aristotle, concepts about the world arise from the common properties of particular things (see figure 2.2). Particular things can serve as models for the family of things that fit the concept or as models for the concept itself. The concept of a horse, for example, is a generalization formed from the observation of a few horses. A plastic figurine in the shape of horse, such as the one being printed in figure 2.3, can serve as a concrete model of a horse. Note that the model need not itself *be* a horse. A concrete model is a physical thing that captures some essence of the things being modeled.

The notion of a concrete model connects naturally to another of Merriam-Webster's definitions, "one who is employed to display clothes or other merchandise." Consider the model in the poster shown in figure 2.4. That model was presumably employed to display something, but interestingly, the merchandise being advertised by this poster, cologne, is not shown on the poster. The model instead was employed to be an exemplar of a "sexy human male" (to be sure, I'm talking about the one in the poster, not the one in the reflection). The model (a human) was employed to serve as a model (an archetype) of a sexy human male. The purpose of such a model is to sell cologne to an individual (perhaps the one in the reflection) by evoking an image of how attractive he might become by wearing such cologne.

Platonic Forms, according to Plato, exist independent of humans, as disembodied truths. A Platonic sphere is perfect. The physical world provides no such sphere. It does not and cannot exist as a physical object. Any physical embodiment of a sphere will be made of some material composed of atoms and molecules. No matter how smoothly polished it is, the surface of the sphere will not match the Platonic concept, but rather will have dents and undulations and a fuzziness imposed by the quantum mechanical impossibility of pinning down the location or boundary of the

Figure 2.2
Plato and Aristotle in a detail of The School of Athens, a fresco by Raffaello Sanzio da Urbino (Raphael) in the Vatican. Aristotle is on the right, gesturing toward the earth, indicating that knowledge arises from the study of things, whereas Plato, on the left, gestures toward the heavens, indicating that knowledge is discovery of Forms that exist in an ideal, disembodied world, independent of humans.

electrons that make up the atoms. Where and what is the surface? Without a surface, we cannot talk about a surface area, but a Platonic sphere should have a surface that is exactly equal to $4\pi r^2$, where r is the radius. But the notion of surface has no rigorous basis in physics at the atomic scale.

So in what sense does the Platonic sphere exist independent of humans? It does not exist in the physical world. We can construct a mathematical *model* of a sphere, but this is still a human construction not a disembodied truth. For example, we can give a mathematical model of a sphere as follows. A sphere with radius r centered at coordinates $(0, 0, 0)$ in a Cartesian coordinate system is the set of points (x, y, z) that satisfy

$$\sqrt{x^2 + y^2 + z^2} = r. \tag{2048}$$

It is no accident that there is nothing sexy about such a Platonic model.

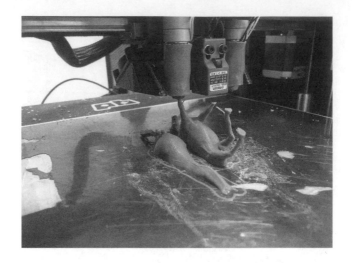

Figure 2.3
A model of a horse being printed by a 3D printer. [Photo by Ben Zhang, courtesy of the photographer.]

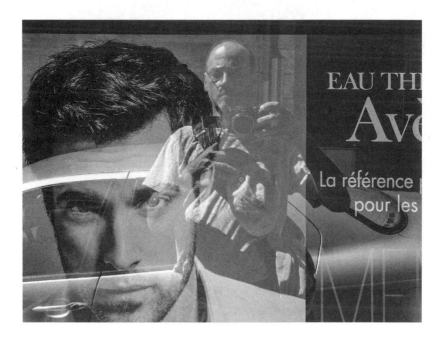

Figure 2.4
Aspirational self-portrait of the author.

The mathematical model of equation (2048) is a human construction, given in the language of algebra and the Cartesian three-dimensional model of space. It can be viewed as an imperfect (i.e., wrong) model of things in the physical world that resemble spheres. Those things in the physical world can be viewed as models of a mental concept of a sphere, as can the equation. But nowhere is there any direct evidence of a human-independent existence of the Platonic Ideal sphere. The mathematical model is not a Platonic Form, but rather at best a *model* of a Platonic Form, existing within a modeling paradigm (algebra and Cartesian space). There are many other ways to model a sphere mathematically, and as I will discuss in chapter 9, every such way has limitations. So even an abstract model is a shadow on the wall. Even if it is just a shadow on the wall, it will be more faithful to a Platonic Form of a sphere than any physical model could be. It may be the best representation accessible to us of the Platonic Ideal.

A concrete model of a horse is shown in figure 2.3 being printed by a 3D printer. That printer accepts as input a file containing another kind of model of a horse. Specifically, the file uses a language called STL (for STereoLithography) that is widely used to specify three-dimensional shapes. The STL language was created in 1987 by 3D Systems, headquartered in Rock Hill, South Carolina, a company that makes and sells 3D printers.

A model of a horse in STL is an abstract model. It is based on a paradigm for modeling three-dimensional shapes in terms of two-dimensional facets that share edges. An example of an extremely simple STL model is shown in figure 2.5. The STL text on the left specifies the pyramid shape at the upper right. It consists of four triangles that form the outer boundary of the object. The vertices of the four triangles are given in a three-dimensional Cartesian coordinate system.

To fully understand the text in the figure, one needs to first assimilate the paradigm on which STL is based. If you will forgive a brief nerd storm, that paradigm is of three-dimensional tessellations of two-dimensional facets, together with rules such as the right-hand rule, which determines which side of a facet is inside versus outside the shape. If you had studied computer graphics, the previous sentence would be easy to read. Otherwise, probably not.

The STL file for the horse being printed in figure 2.3 is much more complicated than the pyramid, so I will not attempt to show it to you here. It is meant to be read by machines not by humans. The horse model is rendered from the STL file at the lower right in figure 2.5, where if you look closely, you can see the two-dimensional facets that define the shape. Such a model of a horse is sometimes called a virtual prototype because it serves the same purpose as a physical prototype, but it does not have a physical form.

Models are expressed in some physical medium. The concrete model of a horse in figure 2.3 is a three-dimensional printed plastic prototype. The physical medium is plastic as assembled by a 3D printer. An abstract model of the same design might be

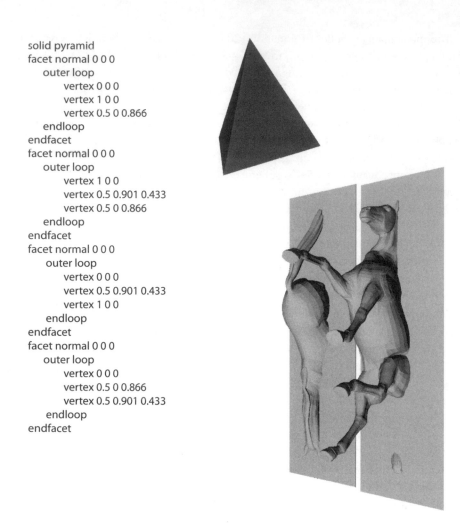

```
solid pyramid
facet normal 0 0 0
    outer loop
        vertex 0 0 0
        vertex 1 0 0
        vertex 0.5 0 0.866
    endloop
endfacet
facet normal 0 0 0
    outer loop
        vertex 1 0 0
        vertex 0.5 0.901 0.433
        vertex 0.5 0 0.866
    endloop
endfacet
facet normal 0 0 0
    outer loop
        vertex 0 0 0
        vertex 0.5 0.901 0.433
        vertex 1 0 0
    endloop
endfacet
facet normal 0 0 0
    outer loop
        vertex 0 0 0
        vertex 0.5 0 0.866
        vertex 0.5 0.901 0.433
    endloop
endfacet
```

Figure 2.5

Three-dimensional shape specified in STL.

a mathematical formula describing its shape, like equation (2048), or an STL file such as that in figure 2.5. This abstract model can be sent to a 3D printer to produce the concrete model. Abstract models also have physical form in the sense that equation (2048) is ink on a page (or pixels on a screen) and an STL file is aligned magnetic iron molecules on a disk or electric charges in a computer memory, but their physical form is incidental. When an abstract model is converted from one physical form to another, for example, into your mental state when you read equation (2048), or when you copy an STL file from one computer to another, we do not end up with two models. It is still just one model, albeit in two or more physical representations. The ontology of an abstract model is independent of its physical embodiment.

Concrete models are exemplars in physical form of a class, whereas abstract models are abstractions of a class. The possibilities for expressive media are much richer for abstract models than for concrete models because they are less constrained by the physical world. STL, in fact, can specify shapes that cannot exist in the physical world, with overlapping facets or facets that do not share edges. Modeling languages are abstract not concrete. They yield more readily to human creativity. Moreover, they invite invention, and even paradigm shifts. A well-chosen modeling language enables elegant expression of complex designs.

Both concrete and abstract models can be used for analysis. A physical prototype of a component, for example, can be used to determine whether the part that it models will fit properly within its housing. But abstract models offer much richer possibilities for analysis. If the modeling medium, the language in which the models are expressed, has a rigorous semantics, then the model may be subject to automated analysis. A computer program can determine whether the component will fit within its housing without ever having to construct a physical prototype.

Figure 2.6 shows a prototype, a concrete model, of an incandescent lightbulb made with a carbonized bamboo filament. This prototype was made in Thomas Edison's lab in Menlo Park, New Jersey. According to The Edison Papers Project (2016), Edison initially tried to make lamps with platinum wire filaments because the metal has a high melting point. But he discovered that when heated in air, the metal would change its structure, weakening the filament, and the melting point would drop. He solved this problem by putting the filament in a vacuum bulb.

Edison was well known for a style of invention that I will call *prototype and test*. To find a material for the filament that would produce a reasonable amount of light with a reasonable voltage and lifespan, Edison tried many alternatives. Although his starting point, platinum, worked reasonably well in a vacuum, platinum is an expensive precious metal. Edison's platinum bulbs were likely to be too expensive to become commercial successes. Also from The Edison Papers Project (2016),

> He turned to carbon and experimented with some cotton threads, different kinds of paper and cardboard, various woods, and then with a few long fiber plant materials before settling on bamboo. Later he had a worldwide search conducted to see if he could find a better long fiber plant as he did not hold the key patents on artificial fibers, which were beginning to prove better.

This is an Aristotelian approach to solving a problem: experiment with materials and infer their properties from observation. The problem he was trying to solve was how to use electricity to generate light. As a side effect of this engineering work, he did some science, discovering a property of the natural world. Specifically, he found that a naturally occurring metal, platinum, when heated in air, changes its structure.

Figure 2.6

Prototype from Thomas Edison's shop of an incandescent lightbulb with a carbon filament. [Image by Terren - Edison Light Bulb, licensed under the Creative Commons Attribution 2.0 Generic license. Lightened by the author. Original from Wikimedia Commons.]

Bamboo filament lightbulbs went into production in 1882 and about six years later were supplanted with tungsten filament bulbs. Both of these styles require operating in a vacuum, otherwise the filament will burn, melt, or otherwise quickly degrade. Edison's discovery that heated metals degrade in air and do not degrade in a vacuum, a scientific fact, became central to the development of a practical lightbulb, an engineering invention.

Thomas Edison used an abstract model of what happens in an incandescent lightbulb, in addition to the physical prototypes. Specifically, he used Ohm's law, first published by Georg Simon Ohm in 1827. Ohm's law relates the current i through a resistor to the voltage v across the resistor by

$$i = v/R, \tag{1024}$$

where the proportionality constant R is called the *resistance*. The resistance, which has units of ohms in honor of Georg Ohm, is a property of the material used to make the resistor and the geometry of the resistor. A lightbulb filament is a resistor, and in Edison's day, the resistance of a filament would have been determined empirically.

Because of its resistance, the filament heats up, and it is because of the heat that the filament generates light. Platinum conducts electricity easily, which means that its resistance is low. Carbon-based materials, such as bamboo fibers, have much higher resistance. At a fixed voltage v, therefore, the current that flows through a platinum filament will be much higher than the current that flows through a bamboo filament. The low resistance of platinum is therefore another disadvantage, along with its high cost. To accommodate the higher currents that result from low resistance, Edison would have had to use thicker copper wires to deliver electricity to the lightbulbs, driving up system cost.

Ohm's law is an abstract model. Unlike the model in figure 2.6, there is nothing physical about this model. It nevertheless represents the "essence of things" and yet, in a true Aristotelian manner, was likely derived by Ohm from observation and measurement rather than from fundamental truths.

Ohm's law can be viewed as a law of nature, in which case it must be true of any electrical circuit. Alternatively, we can view Ohm's law as the *definition* of "resistance" and "resistor." Under the latter interpretation, any device is a resistor if the current that flows through it is proportional to the voltage across it (i.e., if its behavior conforms with the model given in equation [1024]).

The distinction between these two interpretations is subtle but important. First, note that there is an implicit assumption in equation (1024) that we are talking about the current i and voltage v at an instant in time. In almost all electrical circuits, the current and voltage vary with time. For a lightbulb, the current and voltage are both zero when the light switch is turned off and are nonzero when the switch is on, so clearly there is a dependence on time.

A key question now becomes whether the resistance R should also vary with time. It turns out that neither platinum nor bamboo will satisfy equation (1024) with a constant value of resistance. In fact, the resistance varies with the temperature of the material, and the temperature depends on the current. If the filament starts out cold, then as current flows through it, the material will heat up and its resistance will increase. If the filament heats up too much, then the material melts and resistance becomes infinite (no current flows). Hence, the current depends not only on the voltage now, at an instant in time, but also on the history of the voltage and current at earlier times. How long the lightbulb has been on will affect its temperature and hence the current.

So it won't work to fix the resistance R to be constant. We have to let it vary with time. But then we have a conundrum. If R is an empirically determined value, then Ohm's law becomes a tautology! It is trivially true of every electrical circuit. At any instant in time, the resistance is

$$R = v/i, \qquad (512)$$

which is just a rearrangement of equation (1024). Every electrical circuit with nonzero voltage and current trivially satisfies Ohm's law by simply using R as defined in equation (512) as the definition of the resistance. This *defines* the resistance at time t to be whatever value makes Ohm's law true! Surely Georg Ohm did not get a basic electrical unit named after him for discovering a tautology. So this cannot be the right interpretation.

One resolution of this conundrum is that the notion of a *resistor* is a Platonic Ideal Form. A resistor is a device that at all times satisfies equations (1024) and (512) with a constant resistance R. But there is no such device in the physical world. At a minimum, every known material has a resistance that depends on temperature. Moreover, most every known material heats up as current flows through it.[1]

Besides temperature, other physical effects, specifically *inductance* and *capacitance*, ensure that physical materials do not exactly obey Ohm's law with a constant R. These effects introduce memory and dynamics in the system. For example, *inductance* is the tendency for current that is flowing to keep flowing, even if the voltage drops to zero. A material with nonzero inductance will take some time to adjust the current to a new voltage. During that time, it will not satisfy Ohm's law with the same fixed constant R. All materials, in practice, have some nonzero inductance, even if small.[2]

In fact, no physical object is a resistor. Because no physical object obeys Ohm's law, how can we take this to be a law of nature? Plato's allegory of the cave states that *human perception* is limited to shadows of reality, but it appears that *physical objects* are but shadows of the Platonic Ideal Form of a resistor. This Platonic Form is not only inaccessible to humans, it is also inaccessible to nature!

As a consequence, Ohm's law is either trivial or wrong. I see no choice but to conclude that Ohm's law is a human-constructed model, not a fundamental truth about nature. It did not exist as a fundamental truth before Georg Ohm because no physical object in the world obeys it. To the extent that it is a fundamental truth, it is so because we *declare* it to be so. We define a "resistor" to be a physical object whose behavior is reasonably closely modeled by Ohm's equation, and we define

---

1 An exception occurs when some materials, called *superconductors*, are cooled below a critical threshold where they enter a *superconductive* state, where the resistance becomes exactly zero. But the temperatures required are extremely cold. In 1987, Georg Bednorz and K. Alex Müller got the Nobel Prize in Physics for discovering a "high temperature superconductor." Their ceramic compound exhibited superconductivity at the "high" temperature of -243.15 degrees Celsius or -405.67 degrees Fahrenheit. As of this writing, the highest temperature at which superconductivity has been observed is about $-70°C$ ($-94°F$), still extremely cold, and even then only at extremely high pressures. No practical lightbulb could be expected to work only at such temperatures and pressures.

2 An imperfect analogy might help the reader if the reader has not studied electricity. An electric current can be visualized as water flowing down a sluice or channel that is tilted. The degree of tilt is analogous to the voltage. The rate of water flow is analogous to the current. A smaller channel will have a higher resistance than a larger channel (the smaller channel lets through less water for a given tilt). Inductance is analogous to the tendency of water that is flowing to keep flowing (it has inertia). If water is flowing down a tilted sluice and you suddenly flatten the sluice, removing the tilt, the water will not instantly stop flowing. This analogy is imperfect for several reasons. Electric current does not have inertia, or at least not much, and inductance is a property of the channel not the current. But it nevertheless provides a nice visual analogy that can be used to get the basic idea.

Figure 2.7
Drilling through a map. [Photo by Rusi Mchedlishvili, courtesy of the photographer.]

the resistance to be the ratio of voltage across that object to current through that object. An ideal resistor, which does not exist in nature, does not in fact exist at all except in the human mind. Ohm's law was invented, not discovered.

2.3 Models Are Wrong

Modeling is central to every scientific and engineering enterprise. Solomon Wolf Golomb (1932–2016), who has written eloquently about the use of models in science and engineering, emphasizes understanding the distinction between a model and thing being modeled. He famously stated, "You will never strike oil by drilling through the map" (Golomb, 1971). A map is a model. The territory is the thing being modeled. You should drill through the territory, not the map.

For both scientists and engineers, the "thing being modeled" is typically an object, process, or system in the physical world.[3] Let us call the thing being modeled the *target* of the model. The *fidelity* of a model is the degree to which it emulates the target.

---

3 The "thing being modeled" can also be another model. I will examine that issue later in chapter 3.

When the target is a physical object, process, or system, the model fidelity is never perfect. Box and Draper (1987) state, "essentially, all models are wrong, but some are useful." The model in figure 2.4 is not useful (at least not to me). Ohm's law, in contrast, is quite useful. It models certain physical devices, such as Edison's lightbulb filaments. Although it models them imperfectly, Edison used this model to understand that a bamboo filament was a better choice than a platinum filament in a lightbulb.

A useful model has to have a purpose, and the fidelity of the model needs to be evaluated against that purpose. Ohm's law, as a model for a lightbulb, will tell Edison how much current will flow through the filament, but it will not tell him how much light will be generated. A different model is needed for that purpose.

Note that a model may be "useful" in ways that are not practical or mercenary. Merriam-Webster defines "useful" as "helping to do or achieve something." That "something" may be further intellectual inquiry or pure science. That is, a model may be useful because it explains or predicts a phenomenon even if there is no practical application for that phenomenon. Einstein's model of gravitational waves is useful because, among other things, it suggests a way to observe the collision of black holes, as done by LIGO, even if we have no practical use for colliding black holes.

When using models, it is important to apply them only within their regime of applicability, which is limited for all models. Ohm's law, by itself, will not be applicable to a resistor that has melted. Gravitational waves are not useful when studying the interactions of subatomic particles.[4]

Models are generally more useful when their fidelity is higher. So how do we get good model fidelity? We have two different mechanisms available to us. We can either choose (or invent) a model that is faithful to the target, or we can choose (or invent) a target that is faithful to the model. The former is the essence of what a scientist does. The latter is the essence of what an engineer does. Both require assuming that the target is operating within some regime of applicability of the model.

Edison was a quintessential engineer of his time. In selecting a lightbulb filament (a target), among other properties (durability, tolerance for high temperature), Edison needed a filament that was well modeled by Ohm's law. Suppose that Edison had chosen instead a filament that was well modeled by a different law known as Faraday's law of induction. This choice would have resulted in a poor lightbulb. If you will indulge me a brief nerd storm, I will attempt to explain why.

As I pointed out before, *inductance* is the tendency for current that is flowing to keep flowing, even if the voltage drops to zero. An *inductor* is a device that resists *changes* in current, as opposed to a resistor, which simply resists current. By analogy, a resistor is like a lazy person and an inductor is like a stubborn person. It

---

4 Penrose (1989) speculates that gravitational waves may in fact be implicated in certain subatomic quantum mechanical phenomena, but as of this writing, there is no experimental corroboration for this thesis and no wide support among physicists.

takes more effort to get a lazy person to work for you and to keep him working, whereas once a stubborn person is working at something, that person will keep working at it (like me with this book). A person may be both lazy and stubborn, just as a physical device may have both resistance and inductance.

In one of the simpler forms of Faraday's law, the current i and voltage v of an *inductor* are related by[5]

$$v(t) = L\frac{di(t)}{dt}, \tag{256}$$

where the constant L is called the *inductance*.[6] This equation states that the voltage $v(t)$ at time t is proportional to the *rate of change* of current i at time t, where the proportionality constant is L. This means that if the current changes rapidly, the voltage is high. Vice versa, if the voltage is high, the current changes rapidly.

According to Wikipedia,

> Electromagnetic induction was discovered independently by Michael Faraday in 1831 and Joseph Henry in 1832. Faraday was the first to publish the results of his experiments. (Retrieved March 15, 2016)

We might be tempted to change "discovered" to "invented" on the Wikipedia page, but that would not be quite right. The word "discovered" is correct for *induction* but not for equation (256). Equation (256) is an invention. It is an idealized model, and just like Ohm's law, no physical object perfectly obeys it (with constant L). As a model, therefore, it is wrong, but it is extremely useful.

Kuhn (1962) takes a stand on the relationship between discovery and invention, stating, "Discovery and invention are inseparable because the theory to explain the discovery must occur for the discovery to occur." The discovery that a current that is flowing tends to keep flowing (inductance) and that this property is accentuated in certain devices (inductors) is inextricably linked to the model represented in equation (256), in the sense that some form of this model has to be understood to recognize the discovery. This link between discovery and invention, Kuhn says, also makes it much more difficult to pinpoint a discovery, assigning it to a particular person at a particular time:

> $\cdots$ the sentence, "Oxygen was discovered," misleads by suggesting that discovering something is a single simple act assimilable to our usual (and also questionable) concept of seeing. That is why we so readily

---

5 Messerschmitt's law (see footnote on page 2) probably becomes overly conservative when the equation uses calculus, as this one does. I suspect that this equation will drop my readership by more than half, but I will nevertheless stick to the numbering scheme previously established.

6 The units of the inductance L are called "henries" after Joseph Henry. It is customary to use the symbol L for inductance.

assume that discovering, like seeing or touching, should be unequiv-
ocally attributable to an individual and to a moment in time. But the
latter attribution is always impossible, and the former often is as well.
(Kuhn, 1962, p. 55)

Discoveries never occur at an instant in time and are rarely properly attributable to
an individual. The messiness with the discovery of the transistor effect, leading to a
Nobel Prize, years after the transistor had been patented as an invention, underscores
this point.

Let me illustrate how Edison might have used the model of inductance in equation
(256). Suppose that he had chosen as a lightbulb filament an inductor like those in
figure 2.8. The first problem he would have run into is that these filaments would not
have generated any light. But this would only be the start of his problems. Suppose
for simplicity that $L = 1$ henry.[7] Suppose that we now apply a constant voltage of
one volt to the lightbulb. By equation (256), the rate of change of current becomes

$$\frac{di(t)}{dt} = 1. \tag{128}$$

This has units of amps per second. It means that for every second that passes, the
current *increases* by one amp. If the current is initially zero when we turn on the
lightbulb, then after 10 seconds, the current will be 10 amps. After one minute,
the current will be 60 amps. After one hour, the current will be 3,600 amps. After
a few days, the house will have burned down, the transformer on the power pole
outside the house will have blown up, and the electric bill will have become more
than the cost of a college education. Edison would not have been able to sell us more
than one such lightbulb.[8]

I have already concluded that both Ohm's law and Faraday's law are wrong, in
the sense that no physical object obeys either law exactly. But a bamboo fiber in a
vacuum bulb comes pretty close to obeying Ohm's law, and a coil of copper around
an iron core, as on the left in figure 2.8, comes pretty close to obeying Faraday's
law. However, in both cases, the model is wrong.

Edison was an engineer, but he also made contributions to science, and he relied
heavily on experimentation, as many scientists do. To a scientist, the value of a
model lies in how well its properties match those of a target, typically an object

---

7 One henry is actually a very large inductance, but it makes the math simpler and therefore will damage
my readership less than a more reasonable choice of, say, one millihenry.

8 Most household circuits in the United States have fuses that trip, interrupting the current, when the
current exceeds 15 or 20 amps, so this scenario would not play out this way in your house. Also, the
voltage supplied in a household circuit is much larger, typically 170 volts at its peak in the United States
(see Lee and Varaiya [2011] sidebar on page 11 for an explanation of household electric power). Hence,
during the time that the voltage is 170 volts, the current will increase at a rate of 170 amps per second,
which means it will reach 15 amps in 11 milliseconds. At this point, it will trip the fuse, leaving you in
the dark. So the bulb would operate for only 11 milliseconds.

Figure 2.8
A few mostly hand-made inductors. [Image by "me," licensed under CC BY-SA 3.0. Original available at: https://commons.wikimedia.org/w/index.php?curid=1534586.]

found in nature. The value of Ohm's and Faraday's laws lies in how well they describe the properties of some object under study. But to an engineer such as Edison, the value of an object, say a bamboo fiber, lies in how well its properties match a model, in this case, Ohm's law. Edison understood enough about electricity to know that an inductive filament would be of no use. Instead, he knew that he needed a filament for which Ohm's law was a faithful model (and that also generated light), and he went about the task of finding a filament (a target) for that model.

According to Popper's philosophy of science, a scientific model, a "theory," must be falsifiable to be scientific. Under this principle, Ohm's and Faraday's laws are either unscientific or false. If the laws are tautologies, then they are not falsifiable, and if not, then no physical object obeys them, so they are false. Ohm's and Faraday's laws are *useful* not *true*.

In what Simon might have called the "sciences of the natural," to distinguish them from the "sciences of the artificial," a scientist is, by definition, *given* the target. It exists in nature. Such a scientist constructs models to help understand the target. An engineer, in contrast, constructs *targets* to emulate the properties of a model. An engineer uses or invents models for things that *do not exist* and then tries to construct physical objects (targets) for which the models are reasonably faithful. For an engineer, a model provides a *design* and the target is the *implementation*. The task is to find an implementation that is faithful to the model.

These two uses of models are complementary. Engineers and scientists will typically use models *both* ways. Edison spent a great deal of effort characterizing the electrical properties of all sorts of natural materials before settling on bamboo fibers. Good engineering requires doing good science. And at least for experimental science, good science requires doing good engineering, as we saw in the last chapter with the LIGO gravitational wave detector.

In both cases, the models are wrong. In engineering, a model is useful if we can find an implementation that is reasonably faithful to the model. In science, a model is useful if it is reasonably faithful to a target given to us by nature. A scientist asks, "Can I make a model for this thing?" An engineer asks, "Can I make a thing for this model?"

Models are human constructions. Modeling paradigms are also human constructions. Therefore, both are subject to creativity. They are invented not discovered. Because an engineer constructs models for things that do not yet exist, there is much more room for creativity than for a scientist, at least one focusing on the sciences of the natural, who is stuck crafting models for things that already exist. Moreover, I claim that digital technology has smashed open the possibilities for what could exist, so the room for creativity is vast indeed. I examine how digital technology does this in the next chapter.

3 Models of Models of Models of Models of Things

··· in which I argue that in engineering, models are stacked many layers deep, with the design of each layer affecting the designs both above and below it; and that the engineering use of models enables creativity because the layering of models distances designers from the physical constraints of the realization. Digital technology, particularly, has, in effect, mostly removed any meaningful physical constraints from a broad class of engineered systems. Innovation, therefore, is less limited by the physics of the technology than it is by our human imagination and ability to assimilate new paradigms.

3.1 Technological Tapestries

Consider the engineer's question, "Can I make a thing for this model?" Suppose that the answer is "yes" for a broad class of models. For example, technology today gives us the ability to make networks of electrically controlled switches, where closing one switch can cause another switch to open or close. A semiconductor chip is such a network, where the switches are realized as transistors and the network consists of wires that connect the transistors. The medium in which such a network is crafted is the silicon and metal of semiconductors, a physical medium.

Once the answer to the question is "yes, we can make the thing" for networks of switches, then networks of switches become a medium for making models. This medium has its own paradigm, much like Kuhn's scientific paradigms. Just as a scientist uses a paradigm to construct a model of a thing, so does an engineer. The paradigm gives the conceptual framework within which to understand the model.

So what can we build with networks of switches? The network of switches paradigm is quite an expressive one. With just two states for each switch, on and off, it might not seem so expressive, but it turns out that we can interconnect such switches to perform logic functions corresponding to natural language words such as "and," "or," and "not." We can interconnect those logic functions to compare and manipulate strings of bits that represent text and to perform arithmetic on numbers represented in binary. In fact, networks of switches are capable of enormously rich manipulation of any information that is representable as sequences of zeros and ones. It is no accident that transistors functioning as binary switches are the linchpins of information technology.

Once we have the ability to perform arithmetic, we open up the possibility of using another paradigm for design, namely, arithmetic expressions. This paradigm is distinctly different from the network of switches paradigm, but models in the arithmetic expression paradigm are implementable as models in the network of switches paradigm. So if an engineer has a model consisting of arithmetic operations on binary numbers and again asks the question, "Can I make a thing for this model," then again the answer is "yes." To realize the "thing," however, the arithmetic model needs to be first translated into a network of switches model, which then in turn is translated into a silicon chip. Arithmetic expressions become a virtual medium, not directly physical, but translatable into something physical through one level of indirection. This is my first example of transitive models.

It turns out that we can do much more with networks of switches. We can make memory, which stores binary patterns. For example, a bank balance of $256 can be represented by the binary pattern 0000000100000000. There are 16 bits in this representation. It is possible to design a network of 96 switches that can store this number indefinitely. If a customer deposits $16, representable by the binary number 0000000000010000, then a network of switches can add the two numbers, getting 0000000100010000, the binary representation for the number 272. It can then update the memory with the new balance. We can start to see the glimmer of how a computer banking system can emerge from networks of switches.

But thinking about a computer banking system as a network of switches is not practical. For one thing, the number of switches actually required will be vastly more than I've indicated above, and the operations that need to be performed are vastly more complex. A bank will not hire an engineer to wire together transistors, which realize the switches, to make a computer banking system. Instead, the bank will hire an engineer who will write software that will be translated by a computer into a binary pattern that will control a machine that is ultimately composed of a network of transistors. This engineer need not know anything about how to craft a transistor, nor how to perform binary arithmetic using networks of switches, nor how to organize networks of switches to make memories.

In fact, there are many layers of models between the bank engineer and the physical realization. The bank application, a computer program represented as a sequence of letters, numbers, and punctuation, is in fact a model of a model, which in turn is a model of another model, which in turn is yet another model of a model, until ultimately we get down to a model of a thing. Each of these layers of modeling has a paradigm, and each paradigm is a human invention. Only the lowest level physics is given to us by nature.

In this chapter, I will attempt to articulate why such layering of paradigms is so powerful and how the layers turn paradigms into a creative medium that other engineers can use to realize their models. You may come at this with the preconception that these layers of paradigms will be dry, fact-laden technologies, intricate and boring at the same time. But they are not. They are shaped by what is physically possible, but particularly with digital technology, it turns out that so much is possible that they are much more shaped by the personalities and idiosyncrasies of the engineers who create them.

Educators all too often belie the personality of the technology. They present technology as Platonic facts about the world that must be mastered. This is how the educators learned about it. But the creators of the technology did not learn it that way. They invented it, and like literature and art, their inventions reflect the predilections of the creators and the (technological) culture in which they lived. The culture in which they lived was, in turn, defined by the inventions of others. Technology is not a collection of Platonic truths that have always been lurking in the background, waiting to be discovered, but is rather a rich sociological tapestry of ideas created by human inventors. It is shaped by those humans, and had a different set of humans created it, including more women, for example, the technology would unquestionably be different.

I will defer many details to the next two chapters, where I attempt to capture the paradigms and cultures that have manifested in hardware and software technology. In this chapter, I keep a high-level view.

3.2 Complexity Simplified

Engineering of simple systems, like Edison's lightbulb, can be carried out with a prototype-and-test approach. But this approach breaks down as systems get more complex. With more complex systems, the use of models becomes much more important.

Complexity is a difficult concept to pin down. Roughly speaking, something is complex when it strains our human minds to comprehend it. Complexity is therefore a relation between an artifact or a concept and a human observer.

One source of complexity is large numbers of parts. The human brain has difficulty keeping in mind simultaneously more than a few distinct components. In the early days of the telephone network, for example, extensive human studies conducted by Bell Labs determined that people could reliably keep seven numbers in short-term memory but not more. So telephone numbers were constructed with seven digits.

Computers have no such difficulty. They can easily keep billions of numbers "in mind" simultaneously. Computers, therefore, become both a source of complexity for us (we can't understand what they are doing with all those numbers) and a way to help us manage complexity (we delegate to them our memory).

Consider the horse model shown being 3D printed in figure 2.3. The virtual prototype shown in figure 2.5 has more than 23,000 triangular facets. Each facet is specified by nine numbers, so the STL file that defines the virtual prototype contains more than 207,000 numbers represented by more than 46 million bits. Yet my laptop computer generates from these numbers the graphic image shown in the figure, complete with simulated lighting, in less than one second. I can interactively rotate that graphical image to examine all sides of the horse with no noticeable delay for the computer to re-render and re-simulate the lighting at each angle. The rendering of the image requires millions of arithmetic computations on the numbers that represent the vertices of the 23,000 triangles.

It is harder to design a complex system using Edison's prototype-and-test approach because there are so many more possible configurations to try. Nevertheless, prototypes and tests on those prototypes continue to play a major role in engineering today. A modern prototype of an electrical device is shown in figure 3.1 and reported in Choi et al. (2001). This is a transistor of a type called a FinFET, invented at Berkeley by Jeff Bokor, Tsu-Jae King, Chenming Hu, and their students. The prototype shown in the figure, made in 2001, uses the same principles as the field-effect transistor (FET) patented by Julius Lilienfeld (Lilienfeld, 1930).

The innovation in this transistor is its structure, which is more vertical than its predecessors in integrated circuits. Its vertical structure enables many more of these transistors to be packed into a given area of a silicon chip.

I would like to emphasize the dimensions indicated in the figure. The "fin" on the FinFET is 20 nanometers wide. There are one billion nanometers in a meter, so this is quite small indeed.

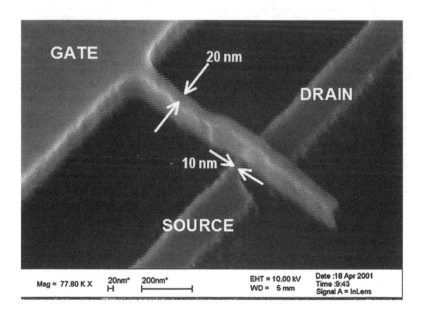

Figure 3.1
Prototype of a modern transistor. [Courtesy of Tsu-Jae King-Liu.]

Consider the implications of being able to realize a transistor that is so small. A modest sized silicon chip is about one centimeter squared. How many 20-nm squares fit in one centimeter squared? Shall I pause for you to do the calculation?

Pause.

OK, hopefully you got the same answer I did, which is 2.5×10^{11}, or 250 billion! This is a square centimeter:

It is hard to imagine fitting 250 billion distinct human-made objects into the space above.

As of 2017, nobody has made a chip with 250 billion transistors (yet), in part, because a chip includes many other things besides transistors, such as wires to connect the transistors. Also, each transistor needs some space around it to separate it from neighboring transistors. So how many transistors can a chip have in practice?

Intel makes a family of microprocessor chips that they call their Haswell line using 22-nm FinFETs. You may have such a chip in your computer. Figure 3.2 shows a portion of a silicon wafer containing several such chips. A "fab" is a high-tech factory that produces such wafers and then cuts them into individual chips and packages them for inclusion in a computer. Each chip in the figure occupies 1.77 centimeters squared, nearly twice as big as the square shown above, and has 1.4 billion transistors (Shimpi, 2013). This is far fewer than 250 billion, but it is still a large number.[1]

Much writing about such technology, including what I've written above, has a breathless enthusiasm about the big numbers. But most of us actually have quite a bit of difficulty assigning any meaning to such numbers because they are so much bigger than anything we encounter in daily life. In fact, the point I want to make is that the human brain is incapable of comprehending any design that has 1.4 billion individual components, each with a potentially different function, despite the fact that the human brain has some 100 billion neurons, each of which does more than a transistor!

Each transistor can function as an electronic switch. It has a control input that either turns the switch on or turns it off. It can turn on and off billions of times

---

1 The particular chip shown in figure 3.2 is a "quad-core + GPU" version of the Haswell product, meaning that each chip actually contains five computers, four "cores" that execute your programs and one "graphics processing unit" that manages the rendering of graphics and text on a screen. The GPU is also a computer, albeit a rather specialized one. If you squint at the figure, you can see the dies for each chip, the rectangular repeating pattern. Within each die, you can see four identical patterns; these are the four cores. The GPU is above the four cores. The rest of the chip is probably mostly memory. As of this writing (August 2016), the largest Haswell chip has 5.56 billion transistors, is about 6.6 centimeters squared, and has 18 cores.

Figure 3.2

Photo of a silicon wafer with several Intel Haswell microprocessor chips. The pin on top is for scale. [Photo by Intel Free Press (Flickr: Haswell Chip), released under a CC BY 2.0 license, via Wikimedia Commons.]

per second. Billions of transistors switching billions of times per second creates unimaginable potential complexity.

How can we design anything using this technology? Can we use Edison's prototype-and-test style of experimentation? Bokor, King, and Hu probably did some prototyping and testing before getting a single FinFET to work. Even so, it was much harder for them than for Edison simply because of the dimensions involved. It is extremely difficult to sculpt a physical structure 20 nm wide. You can't do this with a hammer and chisel. As a consequence, they would have had to make much more use of models than Edison did.

But more to the point, if you want to design a *system* based on a silicon chip, would you start your design by assembling and interconnecting transistors? Consider, for example, the system I am using to write this book. I'm using a software package called LATEX that converts text that I type into a formatted book that can be distributed electronically or printed. Suppose I want to design such a system. Should I start with a bagful of transistors and start connecting them in various ways to see what they do? Most certainly not.

LaTeX is an interesting story. It provides me, a book author, with a paradigm for modeling a book. I construct a model of my book in a text editor that contains annotations such as `\footnote{Footnote contents}` to create a footnote, such as this one.[2] I then run a LaTeX program to convert the text model into a PDF file, another model of pages to be printed. LaTeX was created by Leslie Lamport in the early 1980s, when he was at SRI International. Lamport is an astonishingly prolific and influential computer scientist who received the 2013 Turing Award, sometimes called the Nobel Prize of computer science, for his work on distributed software systems. LaTeX stands for "Lamport's TeX" and is built on top of TeX, designed in the late 1970s by Donald Knuth from Stanford University, another Turing Award winner. Knuth is most well known for his monumental multi-volume work *The Art of Computer Programming*, an encyclopedic compendium of algorithms and principles of programming. Vikram Chandra, in his wonderful book about the aesthetics of software, *Geek Sublime*, said,

> If ever there was a person who fluently spoke the native idiom of machines, it is Knuth, computing's great living sage. (Chandra, 2014)

In an article called "Literate Programming," Knuth argued that software is a literature where code can be written as much to communicate with other human beings as to tell the computer what to do:

> Let us change our traditional attitude to the construction of programs: Instead of imagining that our main task is to instruct a computer what to do, let us concentrate rather on explaining to *human beings* what we want a computer to do. (Knuth, 1984, emphasis in the original)

Knuth created TeX over about 10 years starting in the late 1970s because he found the typography of phototypesetting systems of the day ugly. Today, thousands of people have contributed to TeX and LaTeX, primarily through a system of packages that support an astonishing variety of document preparation needs. It is a thriving, open-source community where nearly all software is free. Almost as if in homage to Knuth, the code gets read and improved by others. The typography that TeX produces, in my opinion, is better than any commercial word processor that I have encountered. In chapter 5, I will have much more to say about the human expressiveness of software.

3.3 Transitivity of Models

A word processing system, such as the one I'm using to write this book, runs on a microprocessor like that in figure 3.2, which uses transistors based on the prototype

---

2 Footnote contents

in figure 3.1. Many levels of modeling exist between the physics of silicon and the word processor. Even more layers can be found between the physics and a system like Wikipedia. Like a pencil, no individual person knows to make such a system. The fact that such systems exist, however, *is* a direct consequence of human ingenuity and creativity. Each layer of modeling allows individuals to contribute to the design without knowledge of or concern for how the layers of modeling they are using came about and without knowledge of how the layers of modeling they are creating will be used by other designers.

A few of the layers involved in the construction of a system such as Wikipedia are shown in figure 3.3. My friend and colleague Alberto Sangiovanni-Vincentelli calls these layers "platforms" (Sangiovanni-Vincentelli, 2007), an apt term because each platform forms a substrate for construction of the models above it. Some of the models above it define platforms for further construction. Sangiovanni-Vincentelli points out that the platforms give designers "freedom from choice." Below a platform there are many possibilities, offering more choices than any human designer can handle. Above the platform, there are fewer choices to be made. You can design more systems by creating a network of transistors than you can using logic gates (explained in chapter 4), for example. But when logic gates provide a suitable platform, the design job becomes much easier if you use logic gates rather than networks of transistors.

Occasionally, one encounters the use of such layers of abstraction in science. But compared with engineering, it is relatively rare, and depth of the layering is much more shallow. Scientists wish to construct models of physical reality, and models of models of physical reality become more suspect simply because they are further from the physical reality.

cloud computing	page 96
libraries, languages, and dialects	page 91
programming languages	page 81
instruction set architectures	page 78
digital machines	page 71
logic diagrams	page 69
logic gates	page 66
digital switches	page 65
semiconductors	page 63

Figure 3.3
Layers of paradigms.

An example from science where layering of models has been successful is the gas laws developed at the end of the eighteenth century. These laws relate pressure, temperature, volume, and mass of a gas, including Boyle's law, Charles' law, Gay-Lussac's law, and Avogadro's law. These models describe phenomena that are ultimately due to the motion of large numbers of molecules in a gas, but they do not describe the phenomena in terms of the individual molecules. For example, Boyle's law states that at a fixed temperature, the pressure of a gas is inversely proportional to the volume it occupies. So, for example, if you reduce the volume (compress the gas), then pressure will increase. These are useful models of models, where the lower level model is of randomly moving molecules colliding with one another and with the surface of the enclosure.

In biology, arguably the most complex of the natural sciences, some researchers have argued that only through such layering can natural biological systems become comprehensible. Fisher et al. (2011), for example, propose the layers shown in figure 3.4 "to tame complexity of living systems." They explicitly propose these layers in analogy to computer hardware systems, even naming some of the layers accordingly, such as "bio-logic gates." The question marks in the figure, however, reveal that this approach is not mature. Biology appears less able to exploit the transitivity of models, compared with engineering, at least so far.

I believe this limitation is quite fundamental. Science cannot benefit as much as engineering from the layering of modeling paradigms. The root of the reason, which I explore more fully in the subsequent chapters, is that engineers build systems to match models rather than models to match systems.

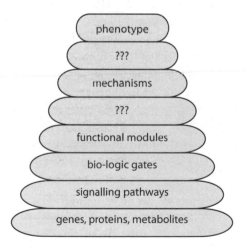

Figure 3.4
Layers of abstraction proposed by Fisher et al. (2011) for synthetic biology.

Even without layering, many phenomena in our physical world (maybe even *most* phenomena) defy scientific modeling. John Searle has written extensively about the inability of scientific models to address cognitive and social phenomena, for example, even though those phenomena are clearly physical. Recall his claim that "the methods of the natural sciences have not given the kind of payoff in the study of human behavior that they have in physics and chemistry" (Searle, 1984, p. 71). His explanation, to my understanding, is a form of failure of transitivity of models. As an illustrative example, he looks at our inability to predict wars and revolutions in terms of lower level physical phenomena:

> Whatever else wars and revolutions are, they involve lots of molecule movements. But that has the consequence that any strict law about wars and revolutions would have to match perfectly with the laws about molecule movements. (Searle, 1984, p. 75)

He points out that we have no laws (in the sense of physical laws) about the occurrence of wars and revolutions, although, ultimately, "wars and revolutions, like everything else, consist of molecule movements."

It is not that higher level phenomena cannot be explained in terms molecule movements. Some can. Searle cites Boyle's law and Charles' law, which can be shown consistent with models of molecule movements. The relationship is relatively simple and the models become predictive. But not so with wars and revolutions. Wars and revolutions are so distant from molecule movements that no such relationship makes sense.

Searle argues that such relationships are *impossible* not just difficult. His reason is quite deep and thought provoking. To make his case, he asks us to consider the concept of "money," which as he points out is "whatever people use and think of as money" (Searle, 1984, p. 78). The fact that this concept is self-referential is a key part of Searle's argument, in that "the concept that names the phenomenon is itself a constituent of the phenomenon." Money can take the form of printed paper, gold coins, or (today) bits stored in a computer and displayed as numbers on a screen. An attempt to explain money as a neurophysiological phenomenon, Searle says, gets tripped up by the many forms that money can take. As we *see* money in these various forms, the stimulus on the visual cortex will be completely different. Searle asks how these completely different stimuli could have the same effect on the brain:

> [F]rom the fact that money can have an indefinite range of physical forms it follows that it can have an indefinite range of stimulus effects on our nervous systems. But since it can have an indefinite range of stimulus patterns on our visual systems, it would ... be a miracle if they all produced exactly the same neurophysiological effect on the brain. (Searle, 1984, p. 80)

So the concept of money must be more than a neurophysiological effect, Searle claims.

> [T]here can't be any systematic connections between the physical and the social or mental properties of the phenomenon. (Searle, 1984, p. 78)

The same argument seems to apply to effects that are quite unlike the sociological concept of money, such as face recognition. We recognize our mother's face in a black-and-white picture of her taken before we were born, for example, despite enormous differences in the physical structure of the face and the material nature of a black-and-white photo versus a real face. It seems that Searle would have to conclude that this too is not a neurophysiological effect. But I suspect it is. The human brain has evolved to categorize visual stimuli into discrete bins despite huge variability in the stimulus.

I'm an engineer, not a philosopher, and not a neuroscientist. I can't credibly reject or defend Searle's argument, but frankly I don't need it to reach essentially the same conclusion. I am perfectly willing to accept that nobody will ever establish any meaningful connection between the physical stimulus to the visual system and the sociological concept of money. Even if we could construct the layers of epiphenomena,[3] their relationships would be so complex, or there would be so many layers, that nothing meaningful could ever arise from their connections. The phenomena at the higher levels are *emergent phenomena*, in that they comprise the lower level phenomena but have their own identity and properties. In later chapters, I will examine the fundamental limits of modeling that make such connections improbable even if the concept of money really is a neurophysiological effect.

But perhaps more interesting, even for some phenomena where we know *exactly* how to explain how they arise from physical effects, it is not useful to do so. In chapter 5, I argue that, although software is ultimately electrons sloshing around in silicon, there are so many layers of modeling between the physics and the software that the connection to the physical is practically meaningless.

I claim that a high-level technology such as Wikipedia has little (and declining) meaningful connection with the underlying physical phenomena in semiconductor physics that make it all work. For digital technology, we can in fact trace the connection from Wikipedia all the way down to semiconductor physics. I will do this for you in chapters 4 and 5. But in doing so, I will show you that there are so many levels of indirection that what happens at the higher level has little meaningful connection with what happens at the lower levels.

Engineers have an advantage over scientists when dealing with layers of models. Natural biological systems and wars and revolutions are givens in our world.

---

3 An epiphenomenon is a phenomenon that can be completely explained in terms of more fundamental phenomena.

Engineered systems are not. For engineered systems, the goal is not to explain them in terms of lower level phenomena. The goal instead is to *design* them using lower level phenomena. This different goal makes it much easier to exploit the transitivity of models.

Consider synthetic biology, which is concerned with designing artificial biological systems. This field is less focused on explaining naturally occurring systems and more focused on leveraging natural biological pathways to synthesize new systems. In synthetic biology, researchers have embraced layered abstractions to great effect. Endy (2005), for example, argues for using predefined functional modules to create biological systems. Indeed, an engineering discipline such as synthetic biology can more readily use layered abstractions because the models need only to model the systems being created. The bioengineers choose the systems to be modeled, and they choose them in part because they *can* model them. To be effective, scientific models need to model the systems given to us by nature, which are much more numerous. And we can't choose those. They are given.

In the next two chapters, I will elaborate on the layers in figure 3.3, with an emphasis on understanding how they came about and with the goal of showing that the specific design of such layers is the creative work of humans, not a collection of God-given facts. But first I would like to spend a little time thinking about how to decide which layer to focus on for any given task.

3.4 Reductionism

At the lowest level, a word processor and Wikipedia are electrons sloshing around in silicon and metal, and the programs that make up Wikipedia are models of models of models of ··· models of electrons sloshing around in silicon and metal. It is tempting to fall into a reductionist trap and say that Wikipedia is "nothing but" electrons sloshing around in silicon, but this would grossly misrepresent reality.

A reductionist perspective explains a system at any level of modeling in terms of the level below it. For example, we could explain how a Wikipedia search uses operators in a programming language that compare text, which are realized by comparisons between binary representations of text in machine code, which uses a compare instruction in an instruction set architecture, which is implemented by microarchitecture with an arithmetic logic unit (ALU) that can do comparison, which is made up of logic gates that implement the comparison, which gates are interconnections of transistors, which transistors are three-dimensional structures of doped silicon. This is a terrible explanation of the search function of Wikipedia.

One of the implications of reductionism is that an epiphenomenon has no effect on the phenomena that explain it. The epiphenomena of temperature and pressure of a gas, for example, can be explained in terms of the underlying molecule movements, but molecule movements would exist unchanged even if we had no concepts of temperature and pressure. But this implication is patently false for the layers of figure 3.3. Only the lowest foundation of these layers, electrons moving an electric field, is given to us by nature. Every other layer is constructed by humankind, often distinctly with an eye toward servicing better the layer above. It is perfectly valid to explain the operation of a logic gate in terms of its role in the design of digital machines and the design of digital machines in terms of the software they are expected to execute. The design of each layer is affected by the layers below *and above* it.

In the sciences of the natural, if scientists were to use such layers, it would be a teleological leap of faith to claim that higher levels of the stack affect lower ones. How could the existence of a biologic gates abstraction in figure 3.4 affect nature's realization of signaling pathways? In contrast, in the stack of figure 3.3, it is not farfetched to claim that transistors are pretty good switches to enable Wikipedia.

In fact, designers of physics-based electronics are constantly trying to improve transistors to make them more like ideal switches. Fundamentally, a transistor is *not* a switch. It is an amplifier. But engineers tune the design of transistors to make them more like switches. For example, when a transistor is off, it is desirable that little current leak through it. This will reduce energy consumption, making it possible to pack more transistors into a small space without generating excessive heat that could melt the silicon. Hence, engineers will tweak the design of the physical structure to reduce leakage. They do this so that Wikipedia can work better. Teleological explanations in this case are perfectly reasonable.

The resemblance, therefore, between the stack of models in figure 3.3 and the one in figure 3.4 is superficial at best. I come back to the point I made in section 2.3, which is that in science, the value of a model lies in how well its properties match those of the target, whereas in engineering, the value of the target lies in how well its properties match those of the model. If our model of a transistor is a switch, then the most valuable transistors are the ones that most perfectly behave like ideal switches.

With sufficient positivist dogmatic determination, we could still insist on a reductionist approach. Once we are *given* transistors by the physical electronics engineers, gates by the VLSI design software, a microarchitecture by Intel, an instruction set architecture by Intel, a Java compiler by Oracle, and a library of Java components by the Eclipse Foundation, *then* we could explain how Wikipedia works in terms of these foundations.

But this is too nerdy even for me. First, these foundations aren't static, so our laboriously constructed explanation could only be valid at an instant. But more important, it vastly understates what Wikipedia really is. At the higher layers of abstraction properties emerge that are difficult if not impossible to explain in terms of the lower level abstractions. An enormous part of the value of Wikipedia lies in its essence as a partnership between technology and culture. I admit a genuine aesthetic delight when I encounter a particularly well-written Wikipedia page and a sense of frustration and gloom when I find a more poorly written page or one that too clearly reflects the views of too few people. A well-written Wikipedia page is difficult to explain in terms of sloshing electrons.

Technology alone does not create a phenomenon such as Wikipedia. Any reductionist explanation of the phenomenon would be naive. In later chapters, I will argue that the failure of reductionism is fundamental and unavoidable in complex technology.

Notice that our layering need not stop at the top of figure 3.3. The software in Wikipedia is created within the modeling paradigms at the top of the figure, but in large part that technology is molded to support a sociological layer above it. But I am a nerd, and I don't understand people, so I won't try to extend my analysis to those sociological levels. I will leave that to the social scientists.

In the next chapter, I will focus on hardware technologies. I point out that hardware does not last nearly as long as the models of the same hardware. Models and the paradigms on which they are based, despite having no material form, are more durable than the things they model, despite those being physical. I focus on digital technology because as we move up from the physical layer (silicon chips), we quickly get extremely expressive media capable of realizing enormously complex and intricate models. The expressiveness of these media unleashes the creativity of humans, enabling the emergence of such transformative technologies as Wikipedia.

In chapter 5, I focus on software technologies. Here, I point out that software encodes the paradigms on which it is constructed. This self-scaffolding enables the bootstrapping of truly innovative artifacts, ones that can profoundly affect human culture. In later chapters, I will explain what software *cannot* do. The door remains open to further creativity.

4   Hardware Is Ephemeral

··· in which I show that hardware is soft, a transient expression of ideas, and those ideas are more durable than the hardware itself. And in which I trace the layered paradigms that make possible digital machines made with billions of transistors.

4.1   Hard and Soft

Steven Connor, professor of modern literature and theory at Birkbeck, University of London, credits the French philosopher Michel Serres, now a Professor of French at Stanford, with developing a subtle and beautiful theory of "the hard and the soft." Serres' thesis, according to Connor, is woven throughout his prolific writings, many of which have not been translated into English. Connor comments that it is difficult to quote Serres, so I will quote Connor instead:

> [T]he contrast between the hard and the soft refers to this distinction between the domain of nature, the object of attention of what we call the "hard sciences," and the domain of culture. The hard means the given, as opposed to the made. It means the physical, as opposed to the conceptual. It means hardware as opposed to software. It means object as opposed to idea, form as opposed to information, world as opposed to word. (Connor, 2009)

Connor finds allusions in Serres' writings to an astonishing array of oppositions between hard and soft, including body and language, science and humanities, things and signs, physical and conceptual, object and idea, form and information, physics and language, a stone and a ghost, motors and information theory, the manual and the digital, sound and meaning, bridge and hyphen, energy and information, flesh and word, the real and the virtual, forces and codes, solids and geometry, objective and subjective, war and religion, a book and a story, or sound and music.

But Serres does not succumb to Platonicity, requiring a sharp delineation between the hard and the soft. Quite the contrary. According to Connor,

> Serres's principal effort is to allow his reader to grasp [the] intermixture [of the hard and the soft.] ... The hard can always evaporate into the soft, the soft calcify into the hard. (Connor, 2009)

In Serres' own words (translated from the French by Connor),

> Hard things display a soft side; material, of course, they engram and programme themselves like software. There is software [*logiciel*] in the hardware [*matériel*]. (Serres, 2003, p. 73)

We then find the more subtle oppositions between the hard and the soft, such as wax and wax, nature and nature, ropes and ropes, or mathematics and mathematics. Each of these, depending on its role and use, can be either hard or soft. "Hardness in softness and softness in hardness," according to Serres.

Following Serres, the three-word summary of this chapter would need to be "hardware and hardware." Hardware is both hard and soft. I will argue that for an engineer who works with digital technology, hardware is merely an ephemeral expression of an idea, lasting longer than a spoken word, which vanishes from the room in a few milliseconds, but still ephemeral, in the grand scheme. The ideas articulated by the hardware, in contrast, although undoubtably mutating and evolving, can last a long time indeed. These ideas are expressed using layers of paradigms that shape and constrain the ideas in ways that even the designer of the hardware is not aware of.

In the rest of this chapter, I outline the layers of modeling used specifically for computer hardware. With apologies to the reader, I confess that the rest of this chapter is a bit of a nerd storm. If you are an impatient reader, or you have no interest in hardware, and you are willing to grant my basic thesis, then please feel free to skip reading the rest of this chapter.

My basic thesis is that the hardware of modern computers is far too complex to design directly. Layers of abstraction are essential, and except for the bottom layer, semiconductor physics, none of these layers is given to us by nature. They are all human constructions, paradigms in Kuhn's sense that frame our thinking about hardware design.

Moreover, I claim that these paradigms, despite having no physical form, are more durable than the hardware. Paradigm shifts are difficult for humans. They can also be quite costly because changes in technology paradigms can mean significant retooling. Software that supports design, such as hardware description languages and their compilers, may have to be redesigned with significant paradigm shifts. Even manufacturing plants may have to change.

Nevertheless, the layering of paradigms makes paradigm shifts easier than they would otherwise be. The design of a microprocessor, for example, often does not need to change when moving to a new semiconductor technology. The emphasis of this chapter is on how this layering of paradigms enables creativity and technological advances. In chapter 6, I will explain how technological advances trigger paradigm shifts.

If you persist in reading this chapter, my primary goal is to show a reader with little or no prior exposure to electronics how the basic operations of an application such as Wikipedia are realized by transistors operating as switches. This explanation cannot be given all at once. It has to be built up in layers. Otherwise, the complexity is simply too much for the human brain. But starting from an abstraction of a single transistor as a switch, we quickly get to abstractions that when realized in a chip require thousands, millions, and billions of transistors. My goal is to show how these layered abstractions enable such scaling up.

4.2 Semiconductors

Modern microprocessors are made from silicon crystals with carefully introduced impurities called dopants. The crystals are sculpted into tiny patterns and shapes like those shown in figure 3.1 in a "fab," a facility where people in bunny suits shepherd silicon wafers through clean rooms, where even the tiniest speck of dust can ruin a chip. The output of a fab is a wafer like that shown in figure 3.2, which is then cut up into individual dies that are inserted into a plastic or ceramic package with metal pins coming out.

When electrical voltages are applied to the chip through the metal pins, currents flow, and models such as Ohm's law, Faraday's law, and quite a few others can be used to understand what happens. A field-effect transistor (FET), for example, uses an electric field to vary the resistance of a "channel" in silicon. When the resistance is low, the transistor is "on." When the resistance is high, the transistor is "off." Because the material may or may not conduct electricity, it is neither an insulator nor a conductor, so it is called a "semiconductor."

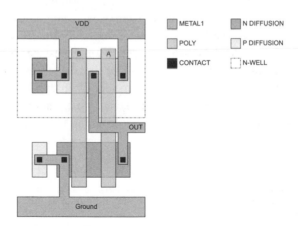

Figure 4.1
Mask design for a four-transistor CMOS NAND gate.

Ultimately, an electrical current is the movement of electrons, and voltages and electric fields arise from electrons piling up in or vacating a location. The behavior of a microprocessor, therefore, is at its root electrons sloshing around in silicon.

Semiconductor physics is a deeply technical specialization that is more science than engineering. In this specialization, the facts of the physical world dominate. Nevertheless, patterns of design have emerged, enabling designers to reuse patterns with rules of thumb to design useful electronic circuits without a deep understanding of the physics. These design patterns constitute a paradigm, and such a paradigm enables an industry to evolve beyond one-off science experiments in the lab.

A chip designer can, in principle, design structures like those in figure 3.1 by carefully specifying how to construct the structure. Such a design takes the form of a set of "masks" that are used in a lithographic process to "print" the chip. An example of a small portion of a mask is shown in figure 4.1. This mask specifies which regions of the silicon crystal should be doped with which dopant, which regions should be covered with polycrystalline silicon, and which regions should be covered with metal. By the late 1970s, when it had become possible to put on the order of 20,000 transistors on a chip, it became impractical to manually design such masks for each circuit.

The layout in figure 4.1 specifies only four transistors. Figure 4.2 shows a die that contains four instances of a similar design to that shown in figure 4.1. That chip has only 16 transistors. The large pads around the periphery of the chip are provided so that wires can be soldered to the chip and connected to metal pins on the exterior of the packaged chip, shown on the right in the figure.

In the late 1970s, Carver Mead of Caltech and Lynn Conway,[1] then at Xerox PARC, wrote a textbook called *Introduction to VLSI System Design*[2] that revolutionized the field by introducing the use of scalable "design rules." Their approach is now universally called the Mead-Conway approach to circuit design.

The design rules define standard layout patterns according to a variable parameter called λ (the Greek letter lambda), which is the minimum distance between features in a layout. The 22-nm Intel process used to fabricate the Haswell chips shown in figure 3.2, for example, has $\lambda = 22 \times 10^{-9}$ meters. With these design rules, chip designers can reuse layouts over and over again, greatly simplifying the design process, even as the feature sizes in chips keep declining.

---

1 To emphasize that paradigms are invented by humans, and sometimes interesting humans, I feel compelled to point out that before working at Xerox Parc, Lynn Conway had been fired by IBM in 1968 after announcing her intention to transition from a male to a female gender role. I cannot even begin to imagine the courage this must have required at that time. She has since become a vocal advocate for transgender rights.

2 The term Very Large Scale Integration (VLSI) started to come into use in the 1970s when chips began to have *thousands* of transistors. The term is still used for chips with *billions* of transistors.

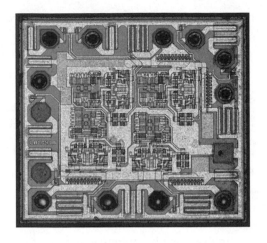

Figure 4.2
Die photo (left) and packaged product photo (right) of an NXP 74AHC00, which contains four 2-input CMOS NAND gates. [Die photo by Mikhail Svarichevsky of ZeptoBars, licensed under Creative Commons Attribution 3.0 Unported License.]

Mead and Conway triggered a paradigm shift, and as with most paradigm shifts, they met with some resistance. Diehard circuit designers insisted that they could design better chips without the constraints imposed by the Mead-Conway approach. They refused to accept this "freedom from choice." Such designers have almost entirely vanished. Today, you may find them in corners of the industry designing specialized chips or performing research on fabrication technologies.

The use of design rules enables a separation between the chip designers and semiconductor physicists. The physicists can specify the design rules, and software can synthesize the masks. This separation also enabled new business models, where chips are fabricated by "silicon foundries," companies such as Taiwan Semiconductor Manufacturing Company (TSMC) or GLOBALFOUNDRIES, that just need to publish their design rules to their customers. Because of the Mead-Conway approach, system design companies can be "fabless," not needing to invest the multiple billions of dollars that it takes to open a fab to make chips. Instead, they contract with the silicon foundries to make their chips.

4.3 Digital Switches

Although transistors can be used for other things (e.g., to amplify a signal), most of the transistors in a microprocessor are used as digital switches. This means that we can understand the behavior of a circuit by understanding whether the transistors are "on" or "off." We don't really have to worry about the behaviors in between.

The following is a standard symbol used to represent a field-effect transistor (FET) like that in figure 3.1:

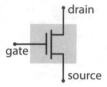

The transistor has a control input wire (called the gate) that is used to turn the transistor on or off. Each transistor has two additional terminals (called the source and drain). When a transistor is "on," we model its behavior as a simple wire connecting the source and the drain. When it is "off," we model its behavior as disconnecting the same two terminals.

The voltage at the gate determines whether the transistor is on or off. For the above transistor, when the gate voltage is sufficiently higher than the source voltage, the transistor is on. Otherwise, it is off.

A "complementary" transistor can also be made, one that is turned on when the gate voltage is sufficiently *lower* than the source voltage. Such a transistor is shown with a circle at the gate:

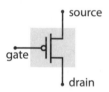

A technology that combines these two types of transistors is called a CMOS technology, for complementary metal oxide semiconductor. The MOS part of this acronym is actually obsolete. It originates from the structure originally used to make these transistors, but the name persists nevertheless. CMOS has ceased to be an acronym in the culture of semiconductor technology. It is simply a noun, pronounced "sea moss."

This switch model of a transistor is an approximation but a particularly useful one. It is a *digital* abstraction, cleanly separating exactly two states, on and off. The real transistor is not so clean: it is electrons sloshing around in silicon. Even good transistors fail to match the digital abstraction perfectly.

4.4 Logic Gates

The digital switch model for a transistor provides a paradigm that we can use to build circuits that perform logic functions. Figure 4.3 shows a circuit diagram for an

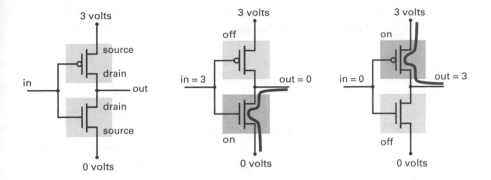

Figure 4.3

Circuit diagram for an inverter logic gate (left). When the input voltage is high (3 volts), the lower transistor is on (center). When the input voltage is low (0 volts), the upper transistor is on (right).

inverter, which converts a high voltage into a low one and, conversely, a low voltage into a high one. If a high voltage represents a digit 1 and a low voltage represents a digit 0, then an inverter converts 1 to 0 and 0 to 1. This is called a "logic gate" because it realizes a simple logical operation: negation.

The circuit is easy to understand even if you have never studied electrical circuits before. The two shaded boxes represent complementary FETs. The top transistor, the one with an extra circle, complements the lower one. In this circuit, the gates of the two transistors are connected together, so when one is off, the other is on.

The figure shows a voltage supplied to the terminal at the bottom of the figure of 0 volts. When the bottom transistor is on, the output will be connected to the 0 volt line, so the output voltage will be 0, as shown in the center diagram. The bottom transistor is on when the gate voltage is 3 volts. So if the input (the gate voltage) is 3 volts, then the output is 0 volts.

The voltage supplied to the top terminal is 3 volts. When the top transistor is on, this 3-volt line is directly connected to the output, so the voltage on the output will be 3 volts. This is illustrated at the right in the figure. The upper transistor is on when the input voltage is 0. So when the input is 0, the output is 3. Hence, the circuit indeed realizes an inverter, assuming that 3 volts represents the binary digit 1 and 0 volts represents the binary digit 0.

The transistor symbols in the diagram represent a simple model for some fairly complicated physics. The circuit is further abstracted to the logic symbol for an inverter:

in ⟶▷∘⟶ out

This represents a digital "logic gate," specifically an inverter. This abstraction has a particularly simple meaning. When the input is the binary digit 1, the output is the binary digit 0 and vice versa.

An inverter is also called a NOT gate because it can be viewed as realizing a logical negation. If the digit 1 represents "true" and the digit 0 represents "false," then the NOT gate converts truth to falsehood and vice versa. When using such a symbol, we no longer explicitly show the supply voltages (connected to the top and bottom terminals). Those are implied.

The connection between switching circuits and logic was apparently first made fully by Claude Shannon, an electrical engineer who will appear again in chapter 7 as the father of information theory. In 1940, Shannon was a 22-year-old master's student at MIT. He wrote a master's thesis that may well be the most influential master's thesis ever written (Shannon, 1940). In his thesis, he showed that electrical switches could be interconnected in a variety of ways to realize any symbolic logic function. For example, one could express with electrical switches a statement such as, "x is true if y is true and z is false or if y is false and z is true." Shannon showed that such logic statements could be designed, analyzed, and optimized using an algebra for logic that had been developed in the previous century by the English mathematician George Boole. Ever since, such circuits have been referred to as Boolean logic circuits. When Shannon was working on his master's thesis, electrical switches were realized using either mechanical relays or vacuum tubes. The transistor hadn't been discovered yet, and although it had already been invented in the 1920s (see chapter 1), it was not widely known or used.

There are a few other useful Boolean logic gates. For example, an AND gate may have two inputs and one output. An AND gate is represented with this symbol:

When both inputs are 1, the output is 1. Otherwise, the output is 0. A NAND gate is equivalent to an AND gate followed by a NOT. It has the following symbol:

An implementation of a NAND gate is shown in figure 4.4; if you now understand how the inverter in figure 4.3 works, then you can probably figure out how the NAND gate works. Shannon designed similar gates using mechanical relays as switches.

An OR gate produces output 1 if *any* input is 1, rather than if *all* inputs are 1, as in an AND gate. An XOR (exclusive OR) gate with two inputs produces 1 if *exactly*

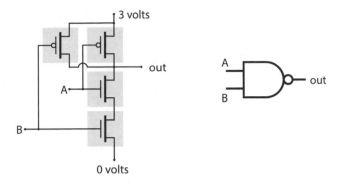

Figure 4.4

Circuit diagram for a NAND logic gate and its logic symbol. When both inputs A and B are high, the output is low. Otherwise, the output is high.

one of the inputs is 1 and the other is 0. In other words, XOR determines whether the inputs differ. I will spare you the implementations of these and even their symbols.

So far, we have three levels of abstraction toward our goal of a Wikipedia system: We have the physical object out of which we will build the system, transistors such as those in figure 3.1, with design rules that free us from having to give detailed geometries for each device; circuit diagrams as in figure 4.3 that abstract transistors as switches; and gates like that at the right in figure 4.3 that abstract the circuit as a logic function. Each of these layers has its own paradigm, with its own lexicon and symbology. But we are still nowhere near being able to build Wikipedia. Let's keep going. I promise we will get there!

4.5 Logic Diagrams

Figure 4.5 shows a logic diagram with 67 logic gates (abstracting roughly 400 transistors that are needed to realize these gates). Some of the gates are inverters like the one in figure 4.3, and the rest are other gates representing logical AND, OR, NAND, NOR, and XOR functions. You need not study this diagram, but rather let's use it to get a sense of this level of abstraction toward the design of Wikipedia.

This diagram specifies the design of an *arithmetic logic unit* (ALU), an important part of any microprocessor. The inputs to this ALU are four-bit binary numbers. They come into the ALU on lines labeled $A_0 \cdots A_3$ and $B_0 \cdots B_3$ at the top. Each input wire carries one bit.[3] There are various outputs at the bottom, including, for

---

3 Shannon credits John Tukey, a mathematician at Princeton and Bell Labs, for the word "bit," a shortening of "binary digit" (Shannon, 1948).

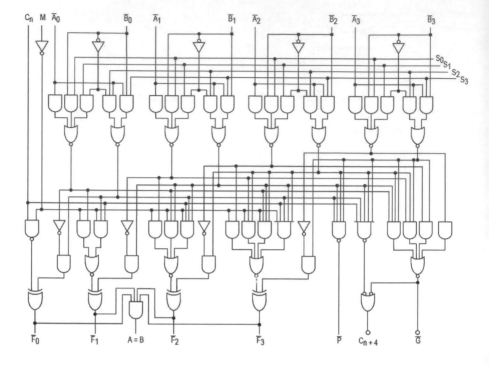

Figure 4.5

Logic diagram for four-bit ALU (Arithmetic Logic Unit). [Image licensed under CC BY-SA 3.0 by Poil, via WikiMedia Commons. From https://commons.wikimedia.org/w/index.php?curid=168473.]

example, one output wire labeled $A = B$. This output tells us whether the two four-bit inputs are equal. Hence, one of the functions performed by an ALU is to compare two numbers for equality. This ALU can also add and subtract two binary numbers, among other functions. In his master's thesis, Shannon showed a simpler but similar adder circuit and showed that each output from such a circuit could be expressed using Boole's symbolic logic algebra.

In a Wikipedia search, each character that I type in a search box to look up a topic is represented by a number, typically an 8- or 16-bit number, not a 4-bit number, so we will need a bigger ALU. But the bigger ALU follows the same principles as this 4-bit ALU, and its logic diagram would be far more intimidating.

To find a page that satisfies my search, Wikipedia needs to compare numbers for equality. The search mechanism is more sophisticated than just comparing numbers, but without being able to compare numbers, it would not be possible to do the search. So we have the first real connection between the hardware and the application, albeit still a tenuous one. We still have a ways to go.

Notice that only by spanning the four levels of abstraction that we have looked at so far can we understand why the world of computers is so obsessed with binary numbers. The reason is simple. Transistors are either on or off. Two states, two numbers, 0 and 1, false and true. That's all we've got, so we have to work with it. There is no 2.

A typical microprocessor today has a much more complicated ALU than this. Today's microprocessors operate on 32- or 64-bit numbers, not 4-bit numbers, so the size of this circuit will be at least 16 times bigger, comprising perhaps 1,000 gates and 6,400 transistors instead of 67 and 400. Typically it is much larger than that, offering many more functions. Clearly a model like that in figure 4.5 becomes unwieldy. You are probably now quite grateful that I didn't show you the logic diagram for the ALU in the laptop computer that I'm using to write this book. It would look similar but with many more gates.

Note that when a logic diagram like that in figure 4.5 is first created by an engineer, although it is a model, no physical realization exists of which it is a model. This underscores the point made in chapter 2 that the model serves a different purpose than a typical scientific model. It can't be true that the value of the model lies in how well it matches the physical system that it models because that physical system does not yet exist. The value of the model *does* depend, however, on our ability to *construct* a physical system whose behavior will match the model. Indeed, the magic of digital technology is that we know how to construct circuits that can match the logic of the model figure 4.5 with almost perfect fidelity, performing the specified operations a billion times per second for years!

4.6 Digital Machines

The ALU of figure 4.5 is just one of many pieces of a microprocessor. If a 32- or 64-bit ALU by itself is too complex to represent with a logic diagram, then certainly a microprocessor will also be too complex. So how does an engineer design a microprocessor?

An ALU can be further abstracted into a single component, as it is in figure 4.6. Near the middle of the diagram is a funny-shaped box labeled "ALU" (or more accurately, labeled "$\underset{\mathsf{A}}{\mathsf{L}}$") that represents a 32- or 64-bit version of the logic diagram shown in figure 4.5. The other boxes in the diagram similarly represent complex logic with many thousands or even millions of transistors. This diagram is easily readable to someone trained in the art of computer architecture. That person could tell you what each of the boxes does. I will not attempt to do that here, but instead I will try to explain the overall style of the diagram.

Figure 4.6 represents the heart of a microprocessor, its *central processing unit* (CPU). It shows, left to right, four stages of the execution of a sequence of instructions

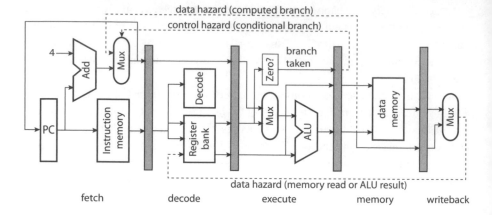

Figure 4.6
Digital machine model of the main pipeline of simple microprocessor (after Patterson and Hennessy, 1996).

that comprise a computer program. At the very left, the components in the diagram fetch instructions from the "instruction memory," where the program is stored (in the form of binary numbers). The decode stage immediately to the right uses logic gates to figure out how to control various parts of the CPU, including the ALU, so that it performs the function demanded by an instruction. For example, if an instruction wants to add two numbers, then the decoder will construct the control input for the ALU to perform addition. The third stage, labeled "execute," uses the ALU to perform the function demanded by the program. The fourth stage, labeled "memory," stores the result of the instruction in memory or uses the result of an instruction as an address for data to fetch from memory.

Figure 4.6 is not a logic diagram. There are no logic gates. Each box represents a component that is realized with many logic gates. The "wires" in this diagram are not simple wires either. The two "wires" into the ALU, for example, represent not one wire, but 32- or 64 wires, depending on whether this is a 32- or 64-bit ALU.

A key idea in the design shown in figure 4.6 is the separation of a memory that stores the program from the logic circuits that execute the program. Earlier computers might have been programmed by actually rewiring the logic circuits, for example, using patch panels with cables and plugs. An architecture with a program memory separated from an ALU, almost universal in computers today, is called a von Neumann architecture, after the Hungarian-American mathematician and computer scientist John von Neumann. Von Neumann first described this style of architecture in an incomplete report titled *First Draft of a Report on the EDVAC*, dated June 30, 1945. Von Neumann had been working on the Manhattan Project, developing the mathematical models for atomic bombs, and was consulting with a project run by the University of Pennsylvania to design a computer called the Electronic Discrete Variable Automatic Computer (EDVAC). The EDVAC was made with vacuum tubes and used a binary (base 2) representation for numbers, unlike its predecessor the ENIAC, which used a decimal (base 10) representation. Although von Neumann is listed as the only author on the report, it appears that many others from Penn had contributed significantly to this design, so referring to this architecture as a von Neumann architecture may once again reflect our need for heroes more than it accurately represents history.[4]

Today, digital designers often assemble designs using hardware components such as ALUs that are reasonably standardized. The ALU shown in figure 4.5 is in fact a standard design called a 74181. Dating back to the 1970s, a series of standard component designs were produced by various manufacturers as a series called the 7400 series, and the four-bit ALU in figure 4.5 was a member of that series. Today, semiconductor manufacturers and computer-aided design (CAD) software provide libraries of standard cells that designers can use to assemble designs. More complex cells are often called simply IP, for intellectual property, and are sold as commodities to designers to use as components in designs.

The tall grey boxes in figure 4.6 represent *latches* or *registers*. A latch is a circuit that, when triggered by the tick of clock, records its inputs. It then holds the input value at its output until the next tick of the clock. In the computer that I am using to write this book, the clock ticks 2.6 billion times per second. At this clock rate, the ALU is presented with an input that lasts only $1/2.6 \times 10^9$ seconds or about 1/3 of a nanosecond. In that 1/3 of a nanosecond, its gates determine the value of its output so that on the next tick of the clock, the result of the ALU computation can be recorded by the downstream latch.

This clocked style of design is known as *synchronous digital logic*. It is *synchronous* in that, conceptually, all the latches are clocked simultaneously. The synchronous digital logic paradigm abstracts away the propagation delays within and between logic gates. As long as the gates are fast enough to correctly perform their function within a clock period, 1/3 of a nanosecond, there is no need to worry about the exact time delay of each gate. The delay can be ignored. The paradigm of logic gates becomes a simple one: they perform their logic function instantaneously.

Figure 4.6 is clearly much more abstract than the logic diagram in figure 4.5. Chip designers call this style of diagram a *register transfer level* (RTL) diagram. I will call it more simply a digital machine because it is a machine operating on words made up of bits. The diagram describes the structure of the microprocessor at the level of operations performed on 32- or 64-bit words and numbers that are exchanged between relatively complicated components. Today's CPUs are actually

---

4 For a wonderful chronicle of von Neumann's pioneering contributions to computation, see George Dyson's 2012 book, *Turing's Cathedral* (Dyson, 2012).

quite a lot more complex than the one shown in figure 4.6. The Intel Haswell line of processors, for example, has up to 19 stages of latches rather than the four shown in figure 4.6. Moreover, these CPUs get combined with complicated hierarchical memory systems. Then they get assembled into servers that contain several CPUs. Then those get assembled into data centers with many thousands of servers and elaborate networks. Only then do we have the hardware for a system such as Wikipedia.

We now have four levels of abstraction, but we still don't have a word processor, much less a system such as Wikipedia. All we have is the *hardware*. We still have to figure out how to tell the hardware what to do. At this point, we need to make a transition from the world of hardware to the world of software. There will be several more levels of abstraction on the software side. I will discuss those in the next chapter. Bear with me. We will get there.

Notice that all four levels of abstraction have existed essentially in their current form for decades. But it would be hard to find any physical piece of such hardware that is several decades old, except in a museum. The abstractions are more durable than the hardware.

The layering of the abstractions is essential to technological progress. When King, Bokor, and Hu introduced the FinFET in 2001, and when it went into production in 2014, the only effect on the other layers is that now they had more transistors to work with. No paradigm change was needed at the upper levels because a FinFET, like previous transistors, makes a pretty good switch. It's just smaller and faster, and it uses less power. This creates *opportunities* at the upper levels, and as I will explain in chapter 6, these opportunities can in fact trigger a crisis that will result in a paradigm shift. But such a "crisis of opportunity" does not *demand* a change at the upper levels. It just enables one.

5 Software Endures

··· in which I argue that the layers of paradigms for software are so deep that the physical world largely becomes irrelevant; that software reflects the personalities and idiosyncrasies of its creators; that software endures much better than hardware in no small part because it encodes its own layered paradigms; and that connected machines in server farms dream.

5.1 Self-Scaffolding

In principle, engineers who wish to write programs could set individual bits in the program memory to get the hardware to do their bidding. In practice, this would be mind-numbingly tedious. A typical program might consist of, say, 10 million bits stored in program memory. That's a lot of zeros and ones. No human could write those zeros and ones without making many, many errors.

The bit patterns that constitute a program are called *machine code*. They are meant for the machine to read, not for the human designer to write. So how do they get written? How does a human designer build a program like a word processor or a system like Wikipedia? Here's where the next stack of abstractions comes into play, those focused on software.

A modern computer program will typically consist of hundreds or thousands of "modules" or "packages," each of which consists of dozens or hundreds of "classes," each of which has dozens or hundreds of "methods," each of which has dozens or hundreds of lines of code. The lines of code are written in a programming language that is translated by a compiler into machine code (often indirectly, first translating into another language, then another language, then to machine code). These layers of design are essential to being able to form any sort of understanding of the behavior of the program.

Way back in 1972, Edsger Dijkstra, a Dutch computer scientist, then a mathematics professor at the Eindhoven University of Technology in The Netherlands, described software as a "hierarchical system," which he defined by analogy,

> We understand walls in terms of bricks, bricks in terms of crystals, crystals in terms of molecules, etc. (Dijkstra, 1972)

He then observed that the number of levels in such hierarchical abstractions is small unless the "ratio between the largest and the smallest grain" is large. The number of molecules in a wall is very large, and yet Dijkstra gives us only four layers.

For the structure of programs, the ratio between a bit of machine code and a program may easily be in the millions. However, in addition, there is layering in the temporal behavior of a program. An individual line of code may execute in a few nanoseconds, whereas the overall function of the program may span hours, days, or months. Dijkstra comments on this ratio:

> I do not know of any other technology covering a ratio of 10^{10} or more: the computer, by virtue of its fantastic speed, seems to be the first to provide us with an environment where highly hierarchical artifacts are both possible and necessary. (Dijkstra, 1972)

To assess Dijkstra's "ratio between the largest and the smallest grain" for software, we need to combine three effects. A computer program may have a million lines of code and get translated into a few million bits of machine code. The machine code is executed on a chip that has billions of transistors, each acting as a switch. And these switches are switching billions of times per second. If the smallest grain is the switching of a transistor and the largest is the computer program, then the ratio is at least 10^{24} or

$$1,000,000,000,000,000,000,000,000.$$

This number is large for something made by humans. Hence, by the time we get to software, we are quite distant from the physical world.

At this point, it ceases to be useful to think of software as a physical phenomenon. It is instead what Hal Abelson and Gerry Sussman call "procedural epistemology" in their introductory computer science book, *Structure and Interpretation of Computer Programs* (Abelson and Sussman, 1996). Software becomes an abstract medium for human creativity and craftsmanship and starts to more closely resemble Searle's cognitive phenomena than the physical phenomena out of which it originates. Software becomes a medium for human expression, not just technical, but also cultural, literary, and artistic. It is of course ultimately realized in the physical world by electrons sloshing around. In words that Connor calls "startlingly sacramental," Serres invokes the Bible when talking about the coalescence of the soft model of software and hard matter that it runs on:

Et verbum caro factum est. (Serres, 2001, p. 78)

And the word was made flesh.

With such a huge difference in scale of the largest and smallest grain sizes, layers of modeling become essential. Unlike the scientific effort that Searle criticizes to explain preexisting sociological phenomena such as money in terms of neurophysiology, our enterprise here is engineering not science. We only need to explain the phenomena we construct, not ones given to us by nature. As humans, we construct software and all the layers below, down to the transistors. It is much easier to explain a phenomenon that we have constructed in terms of a lower level phenomenon that we have also constructed, particularly because the lower level phenomenon was constructed in part to support the upper level one.

Each layer of modeling is governed by a paradigm. A programming language, for example, is such a paradigm. It shapes the programmer's thinking and provides the framework for procedural epistemology. The programming language is a human invention and often reflects the creativity and idiosyncrasies of its inventors.

As a paradigm, a programming language has an interesting property. Specifically, the language can be used, and often is used, to encode its own paradigm. Specifically, a program is translated into a lower level language, such as machine code, by a compiler. Assuming that the machine code is well defined, the compiler encodes the meaning of the programming language and hence encodes its paradigm. But the compiler can usually be written in the very language that it compiles! In fact, it is a common litmus test for a language to be deemed worthy that it be able to encode its own compiler. According to Wikipedia, at least the following languages have compilers written in their own language: BASIC, ALGOL, C, D, Pascal, PL/I, Factor, Haskell, Modula-2, Oberon, OCaml, Common Lisp, Scheme, Go, Java, Rust, Python, Scala, Nim, and Eiffel.

This kind of self-scaffolding of paradigms, I believe, is unique to software among other technologies. It seems to approach Searle's description of sociological phenomena, where "the concept that names the phenomenon is itself a constituent of the phenomenon." And it pervades software from the lowest to the highest layer. At the lowest layer, a microprocessor includes a "boot loader," a tiny built-in program that is executed when the microprocessor is first powered up. The term "boot loader" is a reference to bootstrapping, a term that, according to Wikipedia,

> ··· appears to have originated in the early 19th century United States (particularly in the phrase "pull oneself over a fence by one's bootstraps"), to mean an absurdly impossible action ··· [retrieved April 30, 2016]

At a much higher layer in software, I quoted earlier the Wikipedia page on Wikipedia, itself a form of self-scaffolding. At intermediate levels, to reboot an operating system, something all of us have done, is also a reference to bootstrapping. The operating system uses its own services to start the operating system itself.

In the rest of this chapter, I will explain a few of the layers of modeling that we commonly use in software technology. I hope these explanations will be somewhat less nerdy than the hardware layers of the previous chapter in part because they are more idiosyncratic. I will describe these layers not as facts about the world but as inventions by humans. But I again apologize in advance for the brief nerd storms that I have been unable to keep out of this book. As in the previous chapter, I will progress from lower layers to the upper ones.

5.2 Instruction Set Architectures

Fred Brooks, working at IBM in the 1960s, is credited with developing the idea of an instruction set architecture (ISA), which abstracts what the hardware in a computer does. When a computer executes a program, it executes a sequence of instructions. An instruction may, for example, compare two numbers. Another instruction may specify which instruction to execute next based on the outcome of the comparison. The set of instructions that a computer can execute is called, not surprisingly, its "instruction set." Before Brooks, every distinct model of computer had a different instruction set.

In the 1960s, IBM was developing a family of computers called the System/360 family. One of the goals of the System/360 project was to produce a diverse product line of computers that could all execute the same programs. That is, once you had a bit pattern to put into the instruction memory, that same bit pattern would work on an entry-level computer and on a more advanced, more expensive model. This means that multiple distinct hardware designs all need to interpret programs, stored as bit patterns, in the same way. The hardware could vary by executing programs faster or slower or by providing more or less memory, but the basic functionality of the program should be the same on all instances of the hardware.

To accomplish this, Brooks proposed a standardized "architecture," a specification that defines a fixed instruction set and the bit pattern encoding each instruction. The resulting instruction set architecture is called the IBM System/360 ISA.

It's worth an aside to understand how different Fred Brook's world of computers was compared with ours today. A typical IBM 360, the model 25, could be rented for $5,330 a month or purchased for $253,000 in 1968 (equivalent to about $35,800 and $1.7 million in 2016). The model 25 was aimed at users of "small and medium sized computers" (IBM, 1968). In its largest configuration, its main memory contained 48,000 bytes (each byte is 8 bits). By contrast, the main memory on the laptop I am using to write this book has about 16 billion bytes, and the purchase price was about $2,000.

Despite the enormous differences in cost and scale, Brooks' basic idea of an ISA persists almost unchanged to this day. The ISA used in the laptop on which I am typing this text is called the "x86" instruction set. It was originally introduced in 1978 in the Intel 8086 microprocessor, about 10 years after the IBM 360 first appeared. A variant of the 8086 called the Intel 8088 was used in the first IBM PC, introduced in 1981, shown in figure 5.1.

The x86 ISA grew over time, but it grew in a "backward compatible" way, meaning that an Intel 80186, 80286, 80386, 80486, and many other microprocessors could all execute programs that were written for the 8086. The Haswell processors shown in figure 3.2 are also x86 processors.

Figure 5.1

Original IBM Personal Computer, model 5150. [Image licensed under CC BY-SA 3.0 by Ruben de Rijcke. From https://commons.wikimedia.org/w/index.php?curid=9561543.]

The astonishing persistence of the x86 ISA is a real testament to the power of Brooks' idea. Ironically, the hardware becomes transient, disposable after a few years, and the software endures for decades. Even the word "endure" underscores the irony because this word originates from the old French usage, where "dure" means "hard," and to build a "durable" building, one builds it out of hard material such as stone. Yet in computing, software endures much better than hardware.

Let me illustrate how an ISA abstracts the hardware. Suppose, for example, that one of the tasks for the machine whose hardware is depicted in figure 4.6 is to compare two numbers. If the numbers are equal, then it should branch to a different part of the program. If the numbers are unequal, then it should continue executing the sequence of instructions that it is currently executing. This might be part of the search function on Wikipedia, for example, or a search for the occurrence of a word in a text.

Figure 5.2 shows a small segment of a program for an x86 machine. In the figure, each box represents an instruction. The machine executes the instructions one after the other, from top to bottom. The grey boxes represent arbitrary, unspecified instructions. Two specific instructions are shown. The first is

```
cmp     eax, ebx
```

This instruction compares the contents of two registers named "eax" and "ebx." These two registers contain 32-bit numbers; prior to executing this instruction, the

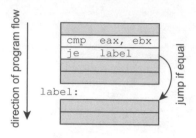

Figure 5.2
Small fragment of x86 assembly code.

program has presumably loaded these registers to contain the numbers representing the characters we are searching for. For example, the characters "Plat" can be loaded into a 32-bit register, assuming that each character is encoded with 8 bits.

The previous instruction is written in assembly language, a textual specification that has to be translated into machine code. The machine code for this instruction is

```
0011010111010100
```

The text in the figure is translated into this binary representation by a program called an "assembler." The "cmp" word is called a "mnemonic" because it is easier to remember than "0011010111010100."

If I may digress briefly, I would like to comment on the culture of programmers. In the 1960s and 1970s, the memory in computers was much smaller than it is today. At that time, it was advantageous to represent an instruction with the mnemonic "cmp" rather than "compare" because "cmp" requires only 24 bits to store, whereas "compare" requires 56. Today, memory is plentiful, but engineers still feel compelled to use short cryptic mnemonics rather than complete words. They will write "fun" rather than "function," "len" rather than "length," and "buf" rather than "buffer." I personally find this an amusing (and sometimes annoying) cultural relic.

The other instruction explicitly shown in figure 5.2 is

```
je     label
```

The mnemonic "je" stands for "jump if equal," and the argument "label" tells the assembler where in the program to jump to if the registers compared by the previous instruction are equal.

There are aspects of this design that seem quite arbitrary, besides the cosmetic choice of mnemonics. For example, why did the designers of the x86 instruction set choose to first compare and then jump in two separate instructions? Why didn't they combine these into a single instruction, such as

```
je     eax, ebx, label
```

They could have done this, but it would have complicated the hardware design because now a single instruction needs to encode four things: the "jump if equal" command, the two registers to compare, and the destination address. The engineering problem that the designers of the x86 architecture were solving straddled two levels of abstraction: the hardware design at the digital machine level, as in figure 4.6, and the instruction set that would be used to specify programs. The phenomenon of assembly language and the lower level phenomenon of the computer hardware affect each other with bidirectional causality.

Engineers who work at these levels are called "computer architects." There is a long and rich history of architectures, most of which have not survived in the marketplace, and some of which operate in very different ways. So-called "dataflow computers," for example, don't even specify programs as sequences of instructions (Arvind et al., 1991). Today, a small handful of instruction set architectures dominate (x86, ARM, SPARC, MIPS, RISC-V, and few others).

Although a computer architect operates at a level quite separated from the "sciences of the natural," ample opportunity exists to use the scientific method to optimize computer architectures. Hennessy and Patterson (1990) revolutionized the field of computer architecture by advocating in their textbook a "quantitative approach," which amounted to systematic use of experiments. A computer architect can form a hypothesis that a particular choice of instruction set design will improve performance and then design experiments to measure the performance on actual programs. This could be done using programs found "in the wild," but these days it is done instead with standardized benchmark suites. The programs in such a suite are idealized models of real programs that attempt to capture their essential features.

The concept of an instruction set architecture brings us safely across the boundary from hardware to software. It is now possible to build applications by writing textual programs. But doing so in assembly language is not a good idea. The programs are too difficult to understand at such a low level of abstraction. To bring up the level of abstraction, computer scientists invented programming languages.

5.3 Programming Languages

Fred Brooks, crediting Aristotle, in a famous paper titled *No Silver Bullet—Essence and Accidents of Software Engineering*, made a distinction between accidental complexity and essential complexity (Brooks, 1987). Essential complexity, Brooks argues, is the complexity inherent in the problem that we are asking the software to solve. Accidental complexity arises from the difficulties that "today attend [software] production but are not inherent."

If I were asked to write this book using a keyboard with only two keys labeled "0" and "1," I could, in principle, do so. The computer, after all, stores this entire book as a sequence of zeros and ones. But it would be difficult to write a book this way. The reality is that this book is difficult enough to write without having to deal with such accidental complexities.

Following Brooks, I would argue that the essential difficulties in writing this book center on how to weave an accessible story around highly technical topics. The technology I have at my disposal has probably removed nearly *all* of the accidental complexities around this task. I have a QWERTY keyboard, and I can touch type quite fast. I have excellent, free, open-source word processing software (LaTeX). When I can't recall what exactly it is that Fred Brooks said in his silver bullet paper, I just go to Google and search for "silver bullet," and I quickly have the paper right in front of me. The only remaining difficulties are the essential ones that follow from the possibly quite controversial cases that I'm trying to make, including the one here, that technology development is a fundamentally creative human activity driven by culture and aesthetics and built on models that are human fabrications much more than discovered natural laws. Only the difficulty of making this case makes writing this book difficult.

The engineered systems that help me, my laptop computer, LaTeX, Wikipedia, and Google, are human constructions of astonishing complexity. The engineers responsible for them relied on tools that also removed many of the accidental complexities, enabling them to focus on the essential complexities. Jimmy Wales and Larry Sanger, who created Wikipedia, did not write their programs in binary or even assembly language. In fact, they used several additional layers of models.

The next layer above ISAs and assembly language is the programming language. In late 1953, John W. Backus at IBM started a project to develop an easier language for expressing programs, particularly those extensively using mathematical expressions. The result of this project was Fortran, a language that endures to this day, with the latest update to the language occurring in 2008.

In Backus' time, "Fortran" was written "FORTRAN." In fact, most everything was written using only capital letters because if you restrict the alphabet to only capital letters, then each letter can be encoded with fewer bits, saving memory. Like the curt mnemonics of assembly code, the use "all caps" became a bit of a cultural relic. My late colleague Chittoor Ramamoorthy, a charming man who went by "Ram" and contributed a great deal to the field of computer architecture, persisted until his death in 2016 in using only capital letters in all his communications, seemingly oblivious to the cultural shift where the use of all caps became yelling.

Figure 5.3 shows a single Fortran statement on a punched card. In the 1950s and 1960s, punched cards were both a storage medium (a stack of cards was a record of the program) and a data entry mechanism. This card reveals one of the key innovations of the Fortran language, which is the use of symbolic variable names rather than memory addresses to refer to numeric quantities that the program is to manipulate. Specifically, the Fortran statement on the card is

```
Z(1)  =  Y  +  W(1)
```

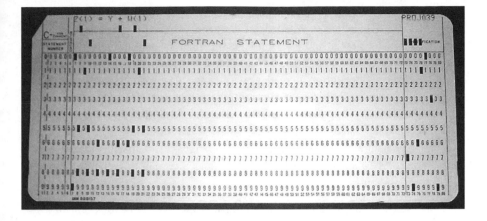

Figure 5.3
Punched card containing one Fortran statement. [Image licensed under CC BY-SA 2.5 by Arnold Reinhold, via Wikimedia Commons. From https://commons.wikimedia.org/wiki/File:FortranCardPROJ 039.agr.jpg.]

Here, Y refers to a value that has presumably been previously assigned, perhaps using a Fortran statement such as

```
Y = 42
```

Z(1) and W(1) refer to the first values in arrays named Z and W.[1] Writing code this way removes the accidental complexity of having to decide where in memory these variables are to be located, choosing registers to temporarily store the values, giving instructions for loading the values from memory into registers, and finally giving the instruction to perform the add. The latter style is what would be required in assembly code.

A Fortran program is translated into assembly code (or directly into machine code) by a compiler. The compiler is responsible for deciding which registers are used for what and where in memory values are stored. Designing good compilers is quite an art, with many opportunities for optimization and experimentation.

But designing good programming languages is more subjective. Programming languages can develop fervent, almost religious followings. Followers of so-called "functional programming," for example, are notorious for their zeal, advocating

---

1 Fortran makes extensive use of arrays as a way to manage memory. For example, an array W with four integers might be declared and then initialized with the statements

```
INTEGER, DIMENSION(4) :: W
W = (/ 42, 43, 44, 45 /)
```

After these statements are executed, the value of W(1) is the integer 42, the value of W(2) is the integer 43, and so on.

languages such as Haskell, named after logician Haskell Curry, and SML, Standard ML. SML is a descendant of ML (MetaLanguage), developed by Robin Milner, a prolific computer scientist and one of my personal heroes who did most of his work at the University of Edinburgh in Scotland and at Cambridge in England. These languages have an elegant mathematical way of specifying computation. Programs can be quite aesthetically pleasing, succinctly stating intent without over-specifying how the intent is to be realized. Nevertheless, among the pantheon of languages, pure functional languages have a small, albeit devoted following.

Much like natural language, programming languages shape the thinking of a programmer. As I mentioned earlier, Abelson and Sussman talk about computing as a "procedural epistemology":

> the study of the structure of knowledge from an imperative point of view, as opposed to the more declarative point of view taken by classical mathematical subjects. (Abelson and Sussman, 1996)

A program is imperative in the sense that it tells a computer what to do, as opposed to a mathematical equation, which tells what is. But there are many ways to tell a computer what to do.

A direct way to tell a computer what to do is to tell it *how* to do it. Computer scientists call a program "imperative" if it specifies a *sequence* of commands, giving a step-by-step procedure, a recipe that the computer is to follow. An imperative program directly represents knowledge as procedure, and like all knowledge, it is surely shaped by language. Most of the widely used programming languages, including Fortran, are imperative languages.

The functional languages, in contrast to imperative languages, adopt the declarative style of mathematics. In an imperative language, for example, the pair of statements

```
x = 1;
x = 42;
```

means to first assign the value 1 to variable x and then to change the value of the variable x to 42.[2] In a declarative language, these two statements are contradictory and will be rejected by a compiler. In a declarative language, the = operator has a different meaning. A statement x = 1 does not *assign* a value to a variable at a particular point in a procedure but rather *declares* that the symbol x *means* 1,

---

2 Among computer scientists, the number 42 is popular to use in examples because of Douglas Adams' *Hitchhiker's Guide to the Galaxy*, a 1978 BBC radio comedy series that later became a "trilogy" of five books. In that story, a special computer called Deep Thought is built to answer the "ultimate question of life, the universe, and everything." The computer takes 7.5 million years to compute and check the answer, which turns out to be 42. The computer reports that the answer seems meaningless because the beings who programmed it never actually knew what the question was.

not at a point in a procedure but always. The order in which such statements are given is irrelevant. In a declarative language, the two statements above are contradictory because x can't mean 1 and also mean 42. Their declarative style is distinctly different from the procedural, step-by-step style of imperative programs. It is perhaps ironic that despite claiming that software constitutes a "procedural epistemology" and "the study of the structure of knowledge from an imperative point of view," Abelson and Sussman's book uses throughout a dialect of Lisp, a functional language originally developed by John McCarthy in the 1950s. Although a Lisp program tells a computer what to do (and hence is imperative in the broader sense of the word), it is a declarative language at its core.

Purely functional languages have been less successful than imperative languages, having only a small but devoted following. As Kuhn says,

> As in political revolutions, so in paradigm choice—there is no standard higher than the assent of the relevant community. (Kuhn, 1962, p. 94)

Kuhn talks about scientific paradigms being incommensurable. They can be irreconcilable accounts of reality, where one paradigm cannot be understood or judged through the conceptual framework and terminology of the other. At their most basic level, programming languages are *not* incommensurable because they are all (today) essentially equivalent to Turing machines, which have an imperative flavor, and to Church's lambda calculus, which has a declarative flavor (see chapter 8). But programmers are usually not using these languages at such an elemental level, and when bundled with the libraries, tools, patterns, and idioms that accompany a language, they arguably do become incommensurable paradigms. Kuhn continues,

> To be accepted as a paradigm, a theory must seem better than its competitors, but it need not, and in fact never does, explain all the facts with which it can be confronted. (Kuhn, 1962, p. 18)

Programming languages do not exist to explain facts but rather to realize algorithms. Any language will realize some algorithms better than others. This may help explain the cacophony of languages that prevail today. If you can indulge me, dear reader, I would like to explore that cacophony.

Wikipedia is realized by an open-source program called MediaWiki, which is written in the PHP programming language. The first version of MediaWiki was created in 2001 by Magnus Manske, then a student at the University of Cologne. His program was later named MediaWiki, a permutation of the name of its biggest user, the Wikimedia Foundation, which runs Wikipedia. As with a lot of open-source software, many people have contributed to MediaWiki since Manske's original design.

Why did Manske use PHP to program MediaWiki? PHP was originally created in 1994 by Rasmus Lerdorf, a Danish-Canadian programmer who worked at

Yahoo. PHP is a "scripting language" specifically designed for building web pages. A scripting language is a programming language intended for specifying short scripts that automate tasks that would otherwise be performed by a human. Notice the layers of culture rather than just technology behind all this: scripting language, wiki, open source, web page. These things did not exist three decades ago, and they are much less new technologies than they are new cultures.

The acronym "PHP" originally stood for Personal Home Page, but according to the current developers, PHP now stands for "PHP: Hypertext Preprocessor." The new name makes PHP a "backronym," which is an acronym where the words are chosen to match the letters rather than vice versa. Moreover, the backronym is self-referential or recursive because the "P" stands for "PHP." Recursion is one of the central tenets of computer science, one of the basic concepts taught in every introductory computer science class. So computer scientists like to pun with recursion.

The use of recursive acronyms was popularized by Richard Stallman with GNU, which stands for "GNU's not Unix!" Choosing a recursive backronym for PHP was, I suspect, a bow to Stallman. GNU is a collection of software that Stallman intended to eventually replace Unix, an operating system originally developed in the 1970s at Bell Labs by Ken Thompson, Dennis Ritchie, and others.

Richard Stallman (figure 5.4) is one of the most influential individuals today in the world of software. Stallman is responsible for a great deal of software that is used

Figure 5.4

Richard Stallman in Vietnam. [Copyright by Richard Stallman, released under "CC-ND," https://stallman.org/photos/.]

worldwide for many purposes. He is also one of the most interesting characters in the story of software. Stallman's GNU project was, in part, a revolt against corporate America. He has spent much of his effort in recent years campaigning against all sorts of encumbrances on software, including software patents, digital rights management, software license agreements, nondisclosure agreements, activation keys, copy restriction, and binary executables that do not include the source code.

In 1985, Stallman launched the Free Software Foundation, which is committed to freeing software. Note my odd use of words; I didn't say it was committed to "free software," which could be easily misunderstood as software that does not cost money. The cost of the software is irrelevant; Stallman uses "free" as in "freedom." In fact, I'm convinced that Stallman anthropomorphizes software, and that his commitment is to freeing the software so the software can go wherever it likes and do whatever it likes, rather than freeing the humans that use software. The latter model, which is better represented by the Berkeley open-source software movement, allows humans to do whatever they like with open-source software. Stallman's model, however, constrains the humans to ensure that they never enslave the software. The copyright notice on GNU software, called the GNU General Public License (GPL), specifically requires that any uses and modifications of GPL'd software preserve the same open rights as the original. This style of copyright is sometimes called a "copyleft" presumably because of seemingly left-leaning politics compared with the right-wing corporate-dominated copyright.

I'm hoping that you can see that the world of software constitutes a diversity of cultures and a literature with parody, social commentary, language, and politics all playing a role, along with technology.

But I digress. My topic in this section is programming languages. Returning to PHP, the language used to create MediaWiki, Lerdorf did not intend PHP to be a new programming language. In an audio interview, Lerdorf noted,

> I don't know how to stop it, there was never any intent to write a prog ramming language $\cdots$ I have absolutely no idea how to write a progra- mming language, I just kept adding the next logical step on the way. (Lerdorf, 2003)

It is not uncommon for major software artifacts to come about this way. They start as small, personal projects, and they grow organically. Lerdorf quipped, "I really don't like programming. I built this tool to program less so that I could just reuse code." With this style of design, the personality and aesthetics of the original authors have a huge impact on the end product.

Besides PHP, several of the most widely used programming languages today, C, C++, C#, and Java, share many essential features with Fortran and with each other. Even so, crossing denominations is rare. A C++ programmer will fight any request to write a Java program and vice versa, often with dogmatic arguments of faith and aesthetics.

In his 1987 Silver Bullet article, Brooks argued that programming languages had advanced sufficiently to have removed nearly all the accidental complexity in programming. The remaining essential complexity, he said, accounted for what has become known as Brooks' law,

Adding manpower to a late software project makes it later.

Brooks first articulated this law in his 1975 book, *The Mythical Man Month* (Brooks, 1975), which is often credited for launching the field called "software engineering." Brooks once quipped that his book is called "The Bible of Software Engineering" because "everybody quotes it, some people read it, and a few people go by it."

But I believe that Brooks vastly underestimated the complexity of systems that were to come, and programming languages alone do not provide an appropriate level of modeling. Many modern software systems consist of millions of lines of code, vastly more than any human can comprehend at the level of the programming language. Instead, programs have to be designed and understood as compositions of subprograms, much the way hardware at the digital machine level abstracts the low-level logic gates.

If you study computer science today, you will likely learn to use only a small number of programming languages, maybe even only one. In my opinion, a software engineer who has mastered only one language is about as well educated as a medieval monk who has studied exactly one book. There is a great deal to be learned from even studying the extensive graveyard of programming languages that have faded from memory.

The following, for example, is a short program in the programming language APL, which I used in a class that I took at Yale in 1978:

$$x[\spadesuit \ x \leftarrow 5?10]$$

APL, which stands for A Programming Language, was developed by Kenneth Iverson in the late 1960s. Iverson received the Turing Award in 1979 for this work. At Yale, an entire computer room was equipped with special terminals that had keyboards that could make the characters in the previous program, such as $\spadesuit$ and $\leftarrow$.

Iverson wanted a language that would concisely specify mathematical operations on entire arrays of data all at once. I can explain the prior program if you can tolerate a short but intense nerd storm. The inner expression 5?10 means, in APL, to create an array with five elements consisting of random numbers in the range 1 to 10 with no repetitions. For example, evaluating this expression in APL might yield the array [9,4,3,8,1]. The expression x $\leftarrow$ 5?10 means that this array should become the value of the variable x. The operator represented by the symbol $\spadesuit$ takes the array that is now the value of x, sorts it, and returns indices that can be used to retrieve the values in the array in numerical order. The construct x[$\cdots$] then uses that array of indices to retrieve the values. So here is a sequence of evaluations that yields a result:

$$x[\mathbin{\triangle}\ x \leftarrow 5?10]$$
$$x[\mathbin{\triangle}\ x \leftarrow [9,4,3,8,1]]$$
$$x[\mathbin{\triangle}\ [9,4,3,8,1]]$$
$$x[[5,3,2,4,1]]$$
$$[1,3,4,8,9]$$

So this program generates an array with five random numbers and then sorts the array so that you get the numbers in increasing order. As you can see, APL programs can be quite cryptic, but they tend to be concise.

A graveyard language that takes the opposite approach is COBOL, after "common business-oriented language." COBOL was designed in 1959 based on an earlier language developed by Grace Hopper, shown in figure 5.5. Hopper was an early proponent of high-level programming languages that were portable, meaning that they could be compiled to execute on a variety of machines, even machines with different instruction set architectures.

COBOL was intended to have a syntax more like English than like mathematics, so it tends to replace symbol operators with words. For example, instead of the APL assignment statement x ← y, in COBOL you would say MOVE y TO x. For many years, COBOL was widely used for business applications such as banking, but today few new programs are written in COBOL.

COBOL and APL represent extremes in an exploration of programming paradigms. COBOL is verbose, using English-language words, with the idea that programs would be more readable by business people. APL is concise, cryptic, and requiring special keyboards. Both succumbed to the Darwinian competition of paradigms, dinosaurs that were once successful but are now largely extinct.

Many more languages are in the graveyard, including Algol, Pascal, PL/I, SNOBOL, Smalltalk, and Prolog. Each of these has interesting ideas and an interesting story. Algol introduced many features present in most modern imperative programming languages, including Java, C, C++, and C#. Pascal introduced the idea of compiling first into a virtual machine language (called byte code) and then executing that program in a program that simulates the virtual machine. This is a centerpiece of the widely used Java language today. SNOBOL, developed at Bell Labs by David Farber, Ralph Griswold, and Ivan Polonsky in the 1960s, introduced high-level manipulation of text, including parsing and pattern matching, a centerpiece of the widely used JavaScript language today, among others. Smalltalk was one of the earliest object-oriented languages, providing a way of structuring programs that is widely used today. Prolog is a "logic programming" language that elegantly expresses rule-based queries over structured data.

Figure 5.5

Rear Admiral Grace Hopper, 1906-1992. Hopper was an early proponent of portable programming languages and pioneered a style of programming where programs read more like English-language sentences than like mathematical expressions. [Image courtesy of the United States Navy.]

Each of these languages encodes a paradigm, a way of thinking about computation. These languages did not die the way Kuhn's scientific paradigms die. No crisis was created by anomalous observations that exposed discrepancies between the paradigm and the natural world. Rather, these languages either mutated into new species of languages (as in ALGOL and Pascal) or progressed toward extinction in a Darwinian competition of survival of the fittest or the most promiscuous (as in APL and COBOL).

5.4 Operating Systems

Today, the word "operating system" means, to most people, one of three things: Apple's OS X, Microsoft's Windows, or Linux. Linux was originally developed in the early 1990s by Linus Torvalds, a Finnish (and later American) software engineer. Linux has become one of the most successful open-source software projects ever, with thousands of contributors and widespread adoption. Linux, like OS X, is based on the Unix operating system originally developed in the 1970s at Bell Labs by Ken Thompson and Dennis Ritchie, with contributions from many others. These three systems, OS X, Windows, and Linux, are the survivors of promiscuous evolution and competition over decades. Today, we should add iOS from Apple and Android from Google, operating systems designed specifically for smartphones and tablet computers.

I could write a great deal about operating systems, but instead I would just like to focus on how an operating system encodes one or more paradigms. A key feature of all these operating systems is the file system. In the hardware of a computer, various forms of memory store sequences of bytes, where each byte is a sequence of eight bits. The laptop computer on which I am writing this book has 16 gigabytes of volatile memory (memory that forgets its contents when you turn off the power) and one terabyte of nonvolatile memory.[3] The nonvolatile memory is sometimes called the "hard disk," although these days it is more likely to be implemented using a type of semiconductor memory called a flash memory rather than the older spinning magnetic disk memory. As far as the hardware is concerned, the contents of the hard disk is just a sequence of a trillion bytes. The hardware can retrieve or update any single byte.

But a list of a trillion bytes, by itself, is not a useful way to organize information. Early operating system designers, such as Thompson and Ritchie, built into the operating system a way of encoding onto these disks a notion of a "file." A file is a subset of the eight trillion bytes that forms a logical unit, and a file system supports assigning a name to the file and organizing files into hierarchical directories, which are also named. With the advent of graphical user interfaces, these directories acquired the metaphor of a "folder," even though, for physical-world folders, folders that contain folders are rather awkward.

My key observation is that nothing in the computer hardware provides the notion of a file and the organization of files into directories. The software embodied in the operating system provides this notion.

The software that realizes the file system is quite clever. It does not even require that the contents of a file be contiguous in memory; the bytes of a file may be scattered all over the disk. The operating system software keeps track of which bytes belong to which file and what directory the file is logically contained in.

Once you have a file system to work with, you no longer need to worry about how your data is stored on a disk. You access the file as a single conceptual unit.

5.5 Libraries, Languages, and Dialects

Millions of lines of code get translated into millions of zeros and ones that coerce billions of transistors to regulate the sloshing of who-knows-how-many electrons.

---

3 Sixteen gigabytes is about 16 billion bytes, and a terabyte is about one trillion bytes.

This is starting to sound like Carl Sagan, whose signature lines involving "billions and billions" frequented his PBS television series *Cosmos: A Personal Voyage* in the 1980s. Sagan was talking about stars and galaxies and often emphasized the incomprehensible range of possibilities, including extraterrestrial life, that such numbers imply.

Digital technology seems to have hit a threshold where the possibilities are limited more by the human imagination than by physical constraints imposed on us by the world. What *can* be accomplished by software far exceeds what we accomplish today, even without further technological improvements. Software has become a digital medium for creativity. I will explore this issue in more detail in chapter 6, and what we *cannot* do with software in chapter 8, but for now let's just focus on how to manage the vastness of the possibilities.

Modern programming languages do, as Brooks claimed, mitigate a great deal of accidental complexity, but not enough to build really interesting systems like Wikipedia. Just as digital machine designs are augmented with standard cell and IP libraries, languages are augmented with component libraries and entire subsystems. As of this writing (August 2016), the Java language, Standard Edition version 8, has 4,240 software components built in for use by software designers. A software engineer will use these components in much the same way that a hardware engineer will use standard cells or an architect premanufactured components such as windows and doors.

Such a library of components becomes like a rich vocabulary, jazzing up the expressiveness of the language. Computer scientists do not usually consider such a library to be part of the language but rather something living and evolving apart from the language. But the library could well have far greater impact on the productivity and creativity of designers than the language. The mechanisms and conventions by which components in the library interact become, in effect, at least a dialect and sometimes even a new language. It is difficult to read a program if you are not familiar with the components it is using, even if you are fluent in its language.

Consider another widely used programming language for web applications, JavaScript, originally developed in 10 days in May 1995 by Brendan Eich. At the time, Eich was working for Netscape, one of the first companies to try to capitalize on the World Wide Web. Netscape was founded as Mosaic Communications Corporation in 1994 by Jim Clark and Marc Andreessen but eventually lost the browser wars to Microsoft and disappeared. Netscape's browser eventually morphed into the widely used, open-source, community-developed Firefox browser.

JavaScript, unlike PHP, is designed to run in the browser rather than in the server. This means that if you access a web page from your laptop or smartphone, and the web page includes a JavaScript program, that program runs on your laptop computer or your smartphone not on the server computer hosting the web page. Many of the web pages you visit routinely include a JavaScript program. Like PHP, the design of the JavaScript language exhibits interesting idiosyncrasies that reflect personal aesthetic decisions made by the original author.

The vast majority of the most widely used websites make extensive use of JavaScript. However, it is difficult to design beautiful and sophisticated web pages with JavaScript alone. Web designers leverage an ecosystem with thousands of "modules" available to designers. Many of these are open-source modules collectively developed by the community, much the way a Wikipedia page is collectively developed.

One widely used JavaScript module for creating sophisticated web pages is jQuery, originally created by John Resig. If you are fluent in JavaScript but do not know anything about the jQuery module, then programs that use it will be unreadable to you. Indulge me to illustrate this.

The JavaScript language, unlike most other programming languages, allows variable names to begin with the dollar sign character, $. The jQuery module defines a single global variable that it calls simply $. That is, the name of the variable is a single character, the dollar sign. This variable gets widely used in programs. To those unfamiliar with this idiom, the program looks cryptic, as a text written in the cyrillic alphabet looks to an English speaker. But the dialect is much richer than implied by just this idiom. Consider the following short JavaScript program:

```
$(document).ready(function(){
    $("#target").text("Hello World");
});
```

If you are fluent in the JavaScript language but unfamiliar with the jQuery module and the modules provided by today's browsers, then this program is completely unreadable. It makes as much sense as the following does to someone fluent in English[4]:

> Twas brillig, and the slithy toves
> Did gyre and gimble in the wabe

Like this poem, the JavaScript program's verse is vaguely familiar but oddly incomprehensible to the JavaScript programmer.

I will spare you the details, but the prior JavaScript program can be used together with an HTML file and a style sheet to create the rather trivial web page shown in figure 5.6. HTML, for HyperText Markup Language, is a completely different language, originally developed in 1980 by Tim Berners-Lee, creator of the World Wide Web, while he was a contractor at CERN, the European Organization for Nuclear Research (the acronym comes from the French name). The HTML that

---

4 From *Through the Looking-Glass, and What Alice Found There*, a novel by Lewis Carroll, 1871.

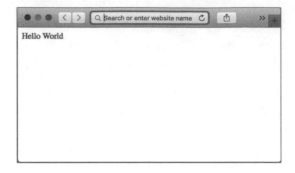

Figure 5.6
Web page using three languages and one dialect to specify.

works with the previous JavaScript program to define the web page in figure 5.6 is as follows:

```
<!DOCTYPE html>
<html>
<body>
    <div id="target"></div>
</body>
</html>
```

Notice the idiosyncratic use of symbols <, >, and /, which were borrowed by Berners-Lee from a documentation format being used internally at CERN at the time.

HTML is universally used today to specify the contents of web pages, along with yet another language called CSS, for Cascading Style Sheets, first proposed in 1994 by Håkon Wium Lie, who was working with Berners-Lee at CERN. To use the previous JavaScript program, HTML is used to define the web page layout, and CSS is used to define the styles used to render the page. For example, if we include the following CSS code:

```
#target {
    color: red;
}
```

then the text "Hello World" will be rendered in red. Notice that the syntax of CSS is quite different from HTML, which is quite different from JavaScript.

The web page of figure 5.6 is constructed using three distinct languages, JavaScript, HTML, and CSS, and one dialect, jQuery, each idiosyncratic and designed largely by a single creative individual. This is perhaps not as culturally rich and diverse as, say, Jerusalem, but it is most certainly not just dispassionate, objective, soul-less technology. It has every element of human subjectivity and invention pervading it. And millions of people today use this particular combination of technologies to design sophisticated web pages.

Of course, we could create a web page like that in figure 5.6 using HTML alone, but there are good reasons for using this combination of technologies. Using JavaScript enables the web page to dynamically update the contents of the page, making it interact with the user. Using CSS separates visual presentation design elements from logical structure and functionality, modularizing the design better. Using jQuery mitigates the accidental complexity associated with the fact that web pages can take a long time (relative to computer speeds) to load from a server and provides convenient access to elements of the page.

Although these languages and dialects each originated with a single individual, all are now thriving open-source communities with hundreds of contributors. They have evolved into a form of collective wisdom, like Wikipedia, rather than individual wisdom, like the Encyclopedia Britannica.

The culture of these communities should make an interesting subject of study for a cultural anthropologist. Resig, for example, first introduced jQuery to the web development community at a conference called a BarCamp in 2006 in New York City. BarCamps might be characterized as the anarchists' conference, in that nobody and everybody organizes the conference. Unlike most professional conferences, which will have a prepublished agenda with all events and presentations defined by an organizing committee, BarCamp participants self-organize using the web, whiteboards, and Post-It notes.

The license history of jQuery also reflects an ongoing passionate debate about the nature of open-source software. It was originally released using a GPL-style license as promoted by Stallman (specifically the Creative Commons CC BY-SA 2.5 license) but was later released under a less restrictive Berkeley-style license called the MIT license.

But I digress again (it is hard to avoid... the background stories are really quite interesting). Let's return to the subject of how to manage the vastness of possibilities that software offers. Software technologies emerge chaotically in a Darwinian ecosystem of ideas. Like a real Darwinian ecosystem, not everyone will agree on what makes one idea more "fit" than another idea, and survival depends more on the ability to propagate than on technical fitness. Promiscuity, personality, money, and culture have enormous, incomprehensible effects.

One approach to understanding this problem is the anthropologists' approach, which is to study the culture as it emerges and attempt to extract wisdom from that study. Just as an anthropologist might use the evolution of natural language as a key part of that study, a software anthropologist might use the evolution of programming languages, with idioms, dialects, and clichés.

One pioneering software anthropology effort was carried out by Erich Gamma, Richard Helm, Ralph Johnson, and John Vlissides in their book on design patterns (Gamma et al., 1994). Widely known as the "gang of four," these authors attempt to categorize a variety of widely used patterns and idioms in software construction. They credit Christopher Alexander, architect, for inspiring their approach. Alexander proposed a pattern language for buildings and cities, and they translated this approach to software. In a testament to the difficulty of this task, in their preface, the gang of four state:

> A word of warning and encouragement: Don't worry if you don't understand this book completely on the first reading. We didn't understand it all on the first writing!

I probably should have included a similar statement in the preface to my book.

The cultural nature of software may help explain why software endures better than hardware. Culture changes much more slowly than technology. The fact that software encodes its own paradigms can also contribute to its durability. For example, although APL is an extinct language, it is easy to find a web page that, using a similar HTML, JavaScript, and CSS combination, presents you with a customized APL keyboard and evaluates any APL programs that you type in.

The cacophony of languages that prevail today is reminiscent of immature scientific fields. Kuhn describes the scientific study of electricity in the first half of the eighteenth century before it acquired its first universally accepted paradigm:

> During that period there were almost as many views about the nature of electricity as there were important electrical experimenters, men like Hauksbee, Gray, Desaguliers, Du Fay, Nollett, Watson, Franklin, and others. (Kuhn, 1962, p. 14)

But even at that time, Kuhn says, these competing paradigms shared a common metaparadigm:

> All their numerous concepts of electricity had something in common— they were partially derived from one or another version of the mechanico-corpuscular philosophy that guided all scientific research of the day. (Kuhn, 1962, p. 14)

The languages used in web technology today, PHP, JavaScript, jQuery, CSS, and HTML, all have a common "mechanico-corpuscular" core, specifically the Turing-Church notion of computation, considered in chapter 8.

5.6   The Cloud

So far we have been talking about individual computers and the software that runs on them. But many of the most interesting uses of computers today are far too big for a single computer to handle. These applications run on computers housed in "server farms," large facilities that can consume up to tens of megawatts of electricity. It is hard to get exact numbers, but various estimates indicate that, as of this writing (2016), Microsoft, Google, and Amazon have on the order of a million servers each in their data centers. Many companies operate perhaps an order of magnitude fewer servers, including Facebook, Yahoo, HP, IBM, eBay, Intel, Rackspace, and Akamai. Each server may contain tens of "cores," individual computers that share certain resources among them, such as memory and network interfaces.

Taken together, this means that there are single software-based services, such as Facebook and Google search, that are running simultaneously on as many as millions of individual computers. These applications make extensive use of PHP, JavaScript, and more general-purpose programming languages like Java, discussed in the previous sections. But they overlay on these languages yet higher level paradigms to handle the distribution of tasks and data across many machines. Languages with curious names such as Pig Latin and frameworks such as ZooKeeper, Sqoop, and Oozie encode the design styles of these paradigms.

For example, Apache Hadoop is an open-source framework that has at its core a distributed file system for spreading data across servers and a realization of a pattern called MapReduce for delegating to servers chunks of a data processing operation. MapReduce was invented in 2004 (and patented) by Jeffrey Dean and Sanjay Ghemawat of Google, although as usual for inventions, the actual novelty is in dispute. MapReduce is strikingly similar to patterns that had existed previously in older software for distributed computing such as MPI (the Message Passing Interface) and database systems.

Hadoop forms an ecosystem of patterns and tools for the design of multiserver applications. Like many of its competitors, Hadoop assumes that hardware failures are common because with millions of servers failure *will* occur. Hardware gets virtualized so that applications can move from machine to machine with minimal disruption. An application may even move from one machine to another machine of an entirely different type, emphasizing the disconnect between the software and the physics of the hardware.

Server applications are often called on to handle truly vast amounts of data, much more than any single computer could handle at once. Consider Google search, for example, which returns in less than a second the results of searching billions of web pages and handles on the order of 40,000 searches per second (Pappas, 2016). How does Google do that? Storing the contents of the web on a computer and searching it each time a query comes in clearly will not work. There is just too much data.

The key to a web search is collecting and indexing the data ahead of time. A major part of the work of Google's servers is not actually responding to search requests but rather reading, indexing, and sorting web pages ahead of time and creating a massive distributed data structure. A "web crawler" finds a website, collects the words on it, and follows the links to other websites, all the while keeping track of link relationships between web pages. The collected data, including statistical features such as word proximity, word ordering, frequency of links, and link associations, are used to build a "memory" of the web that allows for quick recall.

In his book *Turing's Cathedral*, George Dyson draws a thought-provoking analogy:

> The behavior of a search engine, when not actively conducting a search, resembles the activity of a dreaming brain. Associations made while "awake" are retraced and reinforced, while memories gathered while "awake" are replicated and moved around.
>
> . . .
>
> In 1950, Turing asked us to "consider the question, 'Can machines think?' " Machines will dream first. (Dyson, 2012, p. 311)

Indeed, the human brain apparently uses sleep and dreaming to organize information. Does this mean that Google's million-server machines are a nascent intelligence in the process of building some form of cognition by organizing information about the world? I defer this difficult question until chapter 9, where I confront the idea of a digital psyche.

There is quite a bit of sophistication (and secrecy) about how a search works exactly. But one thing is sure: when you perform a Google search, it is not a single computer that replies. Instead, your search is routed through a string of servers depending on the keywords and language patterns in the search so that the search query goes to where the organized data resides. No single server can store and access more than a tiny fraction of the "knowledge" that the servers build while dreaming, so no single computer can reasonably respond to an arbitrary search. The server that first sees your query will be random, based on which servers are available, but after that the query will be forwarded to servers based on keywords and patterns in your search.

Let's consider how much data is involved. First, the amount of data is constantly increasing. The content in the web grows, of course, but more interestingly (and disturbingly, like Orwell's Big Brother), a search engine will watch your every move and use correlations with your previous searches, your location, and even a learned model of your preferences to improve the search results (and to improve the likelihood that advertisements presented to you will be relevant to you). When you read and shop online, you are being read back. The data gathered gets fed into the dream machine to be organized.

It is hard to get solid numbers, but some 2016 estimates suggest that Google may be storing on the order of exabytes of data. One exabyte is 10^{18} or

$$1,000,000,000,000,000,000.$$

That's a lot of data, and potentially all of it can be used to build models of the world, using, for example, the machine learning techniques considered in chapter 11.

The web and users of the web provide a wealth of data for these servers to learn from, but that is not the only source. Software providers are systematically trying to shift all of our computing activity from our personal computers to servers in "the cloud." This changes the business model of software from product to service but, more important, gives the software vendors more direct access to your data, to data about your use of their software, and to data about you.

Even legacy data is being uploaded to these servers. Printed material such as books and journals are being steadily digitized to be added to the online arsenal of information. In 2002, Google started a project to scan every book ever published. Dyson's description of this project is quite extraordinary:

> At the time of my visit, my hosts [at Google] had just begun a project to digitize all the books in the world. Objections were immediately raised, not by the books' authors, who were mostly long dead, but by book lovers who feared that the books might somehow lose their souls. Others objected that copyright would be infringed. Books are strings of code. But they have mysterious properties like strings of DNA. Somehow the author captures a fragment of the universe, unravels it into a one-dimensional sequence, squeezes it through a keyhole, and hopes that a three-dimensional vision emerges in the reader's mind. The translation is never exact. In their combination of mortal, physical embodiment with immortal, disembodied knowledge, books have a life of their own. Are we scanning the books and leaving behind the souls? Or are we scanning the souls and leaving behind the books?
>
> "We are not scanning all those books to be read by people," an engineer revealed to me after lunch. "We are scanning them to be read by an [artificial intelligence]." (Dyson, 2012, p. 312)

Why stop at books? In 2006, Google bought YouTube for $1.6 billion. Because YouTube is a video-sharing website, it provides a truly vast arsenal of data about the world. In 2014, YouTube said that 300 hours of new videos were uploaded to the site each minute, on average. Although the technology for extracting useful information from video and images lags behind that for extracting information from text, we can be sure that the technology will improve. The machines will start to dream in color. As data from sensors comes online, for example, from connected cars, thermostats, and the whole Internet of Things world, what more can the machines learn? I examine this question in the next chapter.

6   Evolution and Revolution

··· in which I argue that technology revolutions differ from scientific revolutions in that
paradigms appear and disappear much more rapidly; new paradigms do not
necessarily replace old ones; and the crises that trigger new paradigms do not arise so
much from the discovery of anomalies but rather from increasing complexity and
technology-driven opportunity.

6.1   Normal Engineering

In his *Structure of Scientific Revolutions*, Thomas Kuhn calls the research that is
firmly grounded in an established paradigm "normal science." The LIGO gravitational wave detector discussed in chapter 1, with its firm grounding in Einstein's
general theory of relativity, despite its monumental scale, qualifies under Kuhn's
scheme as normal science.

Kuhn asserts that adherence to a paradigm is essential to normal science:

> Without commitment to a paradigm there could be no normal science.
> (Kuhn, 1962, p. 100)

He calls normal science "mopping up operations" and "puzzle solving" and asserts
that this is what engages most scientists throughout their careers. The paradigms
within which they operate provide the framework for these operations.

We can similarly define "normal engineering" to be the process of design and
optimization within an established methodology and an established set of rules.
Given a requirement for, say, a web page with some interactive features, a software
engineer is hired to design the HTML and JavaScript code for the web page. This
sort of engineering is easily and effectively outsourced, and a whole industry has
emerged in India to carry out such normal engineering.

Although normal engineering is routine, it nevertheless demands skill and benefits
from talent. When designing a web page, for example, aesthetics are often as
important as functionality. Malcolm McCullough, in his 1996 book *Abstracting
Craft*, focuses on this aspect of normal engineering, observing that digital media,
including the technology for creating web pages and other digital artifacts, offer a
whole new form of craftsmanship. Unlike the physical crafts of, say, pottery and
woodworking, this form of craft works with abstract media, the zeros and ones of
computing. But like the physical crafts, abstract craft admits mastery and aesthetics.

Although normal science certainly admits mastery, it is a real stretch to say it
admits aesthetics. A scientist may object, observing correctly that personal taste is

involved in the selection of experiments to perform, the manner in which they are performed, and the way the results are presented to the scientific community. I have to agree that there is aesthetics in all of this, but the end product of normal science is not an artifact subject to aesthetic judgment. It is, for example, the LIGO validation of Einstein's prediction of gravitational waves. The goal of such validation is not to please the human senses or to stir the soul. It is to reaffirm the Platonic truth of a prevailing paradigm in physics. Kuhn asserts that the object of normal science "is to solve a puzzle for whose very existence the validity of the paradigm must be assumed. Failure to achieve a solution discredits only the scientist and not the theory" (Kuhn, 1962, p. 80). How indeed would LIGO be viewed if it failed to detect any gravitational waves? Would it have undermined Einstein's theory of relativity? Probably not.

A failure to create an effective or successful interactive web page would discredit the software engineers assigned to the task. It would not undermine the paradigm of the web or of the HTML and JavaScript languages. Success in such a project requires some technology, but even more it requires craftsmanship.

Craftsmanship is human skill creating artifacts that did not previously exist. But the craftsmanship in normal engineering is distinctly different from innovation. A beautiful web page that is a pleasure to interact with is not necessarily innovative and almost certainly does not constitute an invention, just as normal science does not seek novelties:

> Normal science does not aim at novelties of fact or theory and, when successful, finds none. (Kuhn, 1962, p. 52)

Craftsmanship and aesthetics can have as much or more impact on the success of an engineering task as innovation. One of the factors in the success of the iPhone is undoubtedly the aesthetic physical design, credited to Jonathan Ive. Amazingly, Apple managed to patent this design, stamping it as an invention. The patent contains one claim, the entire text of which is, "The ornamental design for a portable display device, as shown and described" (Akana et al., 2012). In my opinion, this is an abomination that goes against any reasonable notion of what constitutes an invention. The U.S. Patent and Trademark Office should be ashamed of itself.

Of course, not all of engineering admits aesthetics easily. The design of a sewage handling system for a building usually has only one aesthetic goal: make it invisible. Even so, occasionally even plumbing is used as an aesthetic medium. Witness the Pompidou Center in Paris, which exposes its guts in a bold and aesthetically driven reversal of conventional practice in architecture (see figure 6.1). But with digital media, aesthetic elements are much more common than in other branches of engineering.

As with any craft, mastery of digital media can have an enormous effect on the outcome of a project. But mastery of a craft is quite orthogonal to innovation.

Figure 6.1

The Pompidou Center in Paris exposes the building's mechanical functions for aesthetic reasons. The building was designed by Richard Rogers, Renzo Piano, Gianfranco Franchini, and their teams, and opened in 1977. [Image licensed under CC BY-SA 3.0 by "Reinraum." From https://en.wikipedia.org/w/index.php?curid=37297406.]

Innovation can occur within the framework of an established paradigm, of course. But a truly game-changing innovation, such as the stored-program computer credited to von Neumann or the World Wide Web credited to Berners-Lee, is more like Kuhn's paradigm shifts than like practice within a paradigm. These innovations change the practice of normal engineering for many successor engineers. The question I will address next is what brings about these paradigm shifts. It turns out that the situation in engineering is quite different from that in science.

6.2   Crisis and Failure

Kuhn claims that scientific revolutions occur only after an accretion of anomalous observations made under the old paradigm creates a crisis and only when a new paradigm emerges to replace the old. These are not the forces that drive paradigm shifts in technology.

Paradigm shifts in technology occur for at least three reasons. First, the complexity of systems being engineered overwhelms our human ability to understand or control these systems. For example, programming languages emerged because writing correct machine or assembly code became impossibly difficult. Second, it becomes possible to do something that nobody imagined was possible before. For example,

Figure 6.2

In scientific revolutions, according to Kuhn, new paradigms typically replace old paradigms. In technology revolutions, new paradigms may be built on top of old paradigms, not replacing them so much as hiding them behind a layer of abstraction. (Photos of a sign with the likeness of Charles Darwin on Santa Cruz Island in the Galapagos.)

Google and other search engines enable nearly instantaneous search over everything humans have ever published. Third, complex social, political, and business forces can drive paradigm shifts in technology. Military needs, for example, essentially created aviation, nuclear weapons, and many other technologies, and military budgets provided most of the funding for the early development of computing.

In section 3.2, *Complexity Simplified*, I pointed out that one source of complexity is a large number of parts. Even simple parts with simple functions, such as transistors acting as switches, when there are enough of them, enable enormously complex functionality. Digital technology, rooted in these transistors, has been an enormous source of complexity-driven paradigm shifts for several decades.

In 1965, Gordon Moore, cofounder of Intel,[1] famously predicted that the number of components (transistors, resistors, diodes, and capacitors) in an integrated circuit would double every year for at least the next ten years. In 1975, he revised the

---

1 Moore was one of the "traitorous eight" who left the Shockley Semiconductor Laboratory to found Fairchild Semiconductor and start Silicon Valley.

forecast rate to double approximately every two years. This prediction, widely known as "Moore's law," has been a guiding principle for the semiconductor industry ever since.

In practice, until around 2015, Moore's prediction held steady. The Intel 8080 was a single-chip microcomputer introduced in 1974 with approximately 4,400 transistors. According to Moore's law, therefore, a single-chip microcomputer in 2014 should contain

$$4,400 \times 2^{(2014-1974)/2} \approx 4,610,000,000 \text{ transistors,}$$

which is remarkably close to the 5.56 billion transistors on the Intel Xeon Haswell-E5, introduced in 2014. Although the demise of Moore's law has been predicted many times, most industry observers seem to agree that as of 2015, it has finally significantly slowed.

This rapid acceleration of the capabilities of digital technology has created a steady stream of crises, where inevitably the models and mechanisms used to design and program systems repeatedly break down under the crush of additional capability. As far back as 1972, Edsger Dijkstra, wrote,

> To put it quite bluntly: as long as there were no machines, programming was no problem at all; when we had a few weak computers, programming became a mild problem, and now we have gigantic computers, programming has become an equally gigantic problem. (Dijkstra, 1972)

And that was just 1972! If it was a gigantic problem then, then there is no word for what it is now.

Moore's law refers only to individual computers on individual silicon chips. Today, we find an extraordinary rise in the number of computing devices that are interconnected through networks. Around 1980, Robert Metcalfe, cofounder of 3Com and coinventor of Ethernet, the most widely used wired networking technology today, is said to have postulated what is now known as Metcalfe's law. This law states that the value of a network is proportional to the square of the number of compatible communicating devices on the network. So, for example, if a single isolated device is worth $1, then a network with 10 connected devices is worth

$$\$1 \times 10^2 = \$100.$$

A network with 100 devices would be similarly worth $10,000, and with 1000 devices, $1,000,000. I'll let you calculate Metcalfe's assessment of the worth of the Internet today, which has roughly six billion connected devices.

And the number of connected devices is growing fast. Today, industry leaders breathlessly predict some 50 billion connected devices by 2020, due to the rise of the so-called Internet of Things (IoT). The IoT connects devices that are not first and foremost computers, such as thermostats, cars, door locks, climate control systems, and so on. As they give such predictions, you can almost see the visions of dollars dancing in their eyes.

Because value presumably follows from capability, I assume that Metcalfe would conclude that the *capability* of a network, not just the value, is also proportional to the square of the number of connected devices. Because complexity typically also follows from capability, we should not expect any slowing of crises over the next few years, even if Moore's law grinds to a halt. The crush of increasing complexity does not appear to be slowing.

In any technology, increasing complexity can create crises in two ways. First, designing reliable systems becomes harder. The tools and models that worked well at lower complexity strain until they break as the complexity increases. Second, perhaps as a consequence of the first, the likelihood that a design project will fail increases.

When I worked at Bell Labs in the early 1980s, a large project called AIS/Net 1000 intended to provide bridges between the disparate computing systems that existed at the time. At that time, interconnecting computers through networks was quite a new phenomenon, and as evidenced by Metcalfe's law, the value of such interconnections was recognized, at least by Metcalfe.

But these interconnections exposed many incompatibilities between the computer systems, particularly computer systems from different vendors. These systems had been designed to work in isolation. The binary bit patterns used to represent numbers and text, for example, were different. The order in which bits were arranged in memory differed. The protocols and speeds used to communicate differed. These differences meant that one computer often could not directly communicate with another, and even if it could, it would not interpret the bit patterns produced by the other correctly. In effect, each computer operated within its own paradigm, and the paradigms were incommensurable.

As computers became networked, these incompatibilities triggered a crisis. AIS/Net 1000 aimed to solve this problem by performing translations within the network, permitting the disparate paradigms to persist. When one computer sends a message to another, the message would be automatically translated during transport. Nobody would have to change how they did things, and AT&T would sell the glue that enabled interoperability. This was to be the Babel fish of networks.[2]

---

2 Douglas Adams described the Babel fish in *The Hitchhiker's Guide to the Galaxy*. According to Adams, "The Babel fish is small, yellow, leech-like, and probably the oddest thing in the universe. It feeds on brain wave energy, absorbing all unconscious frequencies and then excreting telepathically a matrix formed from the conscious frequencies and nerve signals picked up from the speech centres of the brain, the practical upshot of which is that if you stick one in your ear, you can instantly understand anything said to you in any form of language: the speech you hear decodes the brain wave matrix."

The project failed. AT&T wrote off more than $1 billion of development effort. It turns out that few customers were actually willing to pay for this service. Instead, the paradigms saw a Darwinian consolidation. Competing species were unable to coexist within the same ecosystem of networked computers.

Instead of a Babel fish, the Internet emerged. A key enabler for the Internet was the acceptance of the so-called Open Systems Interconnection (OSI) model. OSI is a layering of modeling paradigms, sketched in figure 6.3. Like the layers in figure 3.3, each level of the OSI model provides a conceptual framework for communication between computers. The lowest level, called the physical layer, is concerned with transporting sequences of bits from one place to another without concern for what the bits mean. The layers above this assign more meaning to the bits. For example, layer 6, the presentation layer, may treat a collection of, say, one million bits as an encoding of an image in a particular format, such as the standardized JPEG format widely used in digital cameras and on the web.

Kuhn talks about paradigms being incommensurable. In the OSI model, the terms "frame," "packet," "segment," and "session" all refer to a finite collection of bits, but they all have different meanings at different levels. Understanding these different meanings is one of the most confusing parts about working with low-level networking software, in my experience. It is much easier to work at exactly one of these levels and not try to cross layers.

Calling the layers of the OSI model "paradigms" is perhaps a bit odd because they differ significantly from Kuhn's scientific paradigms. Like Kuhn's paradigms, they do provide a mental model for humans to understand how a system operates. For example, it is a different mental model to visualize one computer sending an image, a photograph, to another, versus visualizing one computer sending a stream of one million bits. But unlike Kuhn's paradigms, for these layers to work, their definition must be made absolutely precise. A misinterpretation of a single bit among one million bits may render an image unreadable. Kuhn's paradigms are much more robust; they can tolerate a certain amount of creative misunderstanding, which can sometimes form the engine for innovation or even paradigm shifts.

7. application	network services used directly by applications
6. presentation	interpretation of bit patterns as text, images, numbers, etc.
5. session	multiple back-and-forth data exchanges treated as a unit
4. transport	reliable transmission of data segments
3. network	routing of packets of bits in a multi-node network
2. data link	delivery of a frame of bits between two nodes
1. physical	streams of bits over wires or radio

Figure 6.3
The OSI model for communication between computers.

It is not easy to make the OSI model layers precise. For computers on the Internet to reliably communicate, they all need to agree on precise meanings at every layer, down to the interpretations of each individual bit. The process of building the standards that codify this agreement can be a messy, political, and bureaucratic morass of conflicting national and business interests.

As a case in point, it may be helpful to understand how the OSI model came about. The OSI model is a joint effort of two standardization bodies, the International Organization for Standardization (ISO) and the Telecommunication Standardization Sector of the International Telecommunication Union (ITU-T, formerly CCITT), which had each separately developed similar models for communicating computers in the late 1970s. But similar models are not enough to get computers to communicate. The models have to be identical. Hence, these two bodies got together to publish a joint document, a process that no doubt involved considerable bickering over minutia.

To get a sense of all the competing interests that get involved, it may be helpful to understand how these standardization bodies are organized. The ISO is composed of representatives of various national standardization agencies from some 162 countries. The ITU-T, a United Nations agency, coordinates standards for telecommunications. In addition to representatives from many governments, these bodies include representatives from competing businesses, some of which will have already sunk considerable investments into the technologies being standardized. As a consequence, the battles that can emerge over standards development can be prolonged and painful, and the ensuing compromises can sometimes undermine the effectiveness of the resulting standards.

JPEG, one of the most commonly used encodings for photographs and a level-6 (presentation layer) standard in the OSI model, is an acronym for the Joint Photographic Experts Group, which created the standard. This group is a committee overlapping ISO/IEC JTC1 and ITU-T, the same organizations involved in the OSI model, except for the addition of the International Electrotechnical Commission (IEC). The IEC is a nongovernmental international standards organization that develops standards for electrical, electronic, and related technologies. Surely by now, from the barrage of acronyms, you see how bureaucratic all of this is.

As with many such standards, one of the complexities that arose with JPEG concerned intellectual property. A major challenge in establishing such international standards is to ensure that anyone can legally use the standard without infringing on the rights of someone else. There can be quite a bit of posturing during the development of a standard, where businesses will attempt to ensure that the use of the standard requires license payments to them for patents that they hold or where patents that they hold will give them a competitive advantage when implementing the standard. Organizations can even be quite sneaky about this, concealing their business interests from the standards body until it is too late to change the standard. As a result, standards often do not reflect the best technical solutions to a problem.

In the case of JPEG, after publication of the standard, several companies asserted that the standard infringed on patents that they held. In a collection of notable cases starting around 2007, a patent holding company called Global Patent Holdings, LLC, claimed that the act of downloading a JPEG image from a website or sending it through email infringed a patent that it held, U.S. Patent 5,253,341 by Rozmanith and Berinson (1993). A messy set of suits, countersuits, and threats ensued. According to a Wikipedia article on JPEG,

> Global Patent Holdings had also used the '341 patent to sue or threaten outspoken critics of broad software patents. (https://en. wikipedia.org/wiki/JPEG, retrieved April 26, 2016)

After extensive battles, the U.S. Patent and Trademark Office issued a Reexamination Certificate in 2009 canceling all claims of the patent, asserting that prior art invalidated the claims. By this time, many organizations had wasted enormous amounts of money on completely nonproductive fights over intellectual property.

A patent holding company is a corporation that does not manufacture or sell products but just acquires and holds patents for the purpose of extracting royalty payments from companies that do sell products. Such companies are often called "patent trolls," after the troll in the Norwegian fairy tale "Three Billy Goats Gruff" who eats anyone who tries to cross the bridge under which it lives.

The emergence of patent trolls has significantly changed the business climate for technology companies in the United States. An organization that produces a product and also owns a patent portfolio may be hesitant to sue another organization that also owns a patent portfolio because that other organization may countersue for patent infringement. But an organization that does not produce any products is much less vulnerable to countersuits. These organizations exist only for the purpose of siphoning money from organizations that produce products. In my opinion, they are parasites.

But I digress. The main point is that it is not only creativity that determines the nature of the layers of paradigms in a technology, but that complex business and political interests can also intervene. These layers, therefore, are not in any sense objective truths. They are the result of flawed and human processes.

AIS/Net 1000 failed to solve the crises created when incompatible computers started to become networked. That crisis has since been largely resolved through the emergence of the Internet, which depends on standardization of all levels of the OSI model. The so-called Internet Protocol (IP, not to be confused with intellectual property) is at level 3 in the OSI model, the network layer. All traffic in the Internet uses IP. A layer above, at the transport layer, is the widely used Transmission Control

Protocol (TCP), which overlays on IP the concept of reliable transmission. Specifically, a TCP/IP packet sent to a computer must be acknowledged by that computer. The sending computer will repeatedly send the packet until it receives an acknowledgment. As a consequence, barring catastrophic failure in the network or in the sending or receiving computer, every sent packet is eventually received. TCP also ensures that packets sent in order are received in the same order. TCP is essential to email, among many other services.

Email, in turn, relies on another protocol called SMTP, for Simple Mail Transfer Protocol, a level-5 protocol (session layer). At this level, a sequence of TCP/IP packets is collected into a unit, an email message. The design of this protocol is greatly simplified by being able to assume the properties of the lower layers, specifically that packets are delivered reliably and in order. If you send a JPEG image by email, then the level-6 (presentation layer) protocol for JPEG image encoding defines the interpretation of the bits contained in the packets as an image.

The OSI model provides separation of concerns, where routing of packets, reliable delivery and ordering, email addressing and content, and encoding of the email payload are all separated. Each of the protocols involved is much easier to design and understand because it uses the properties of the layers below, and it avoids providing capabilities that will be provided by a layer above.

AIS/Net 1000 offered a solution to a crisis but it turned out not to be the solution that prevailed. I believe a key reason is the lack of separation of concerns. AIS/Net 1000 was to be the single solution to interconnectedness, whereas the OSI model made it possible for many competing solutions at each layer to fight it out, creating a Darwinian ecosystem with distinct niches within which solutions could compete.

There are other spectacular technology failures with similar reasons. Ed Cone blogs about the failure of the Federal Aviation Administration's (FAA's) Advanced Automation System (AAS) project, conceived in 1981 and terminated in 1994 (Cone, 2002). In this project, the FAA contracted IBM Federal Systems, a division of IBM later acquired by Lockheed Martin, to replace the nation's air traffic control system with a completely new modern design. According to Cone, "the FAA ultimately declared that $1.5 billion worth of hardware and software of the $2.6 billion spent was useless."

The historical backdrop of the project is telling. In 1981, the air traffic controllers went on strike, and President Ronald Reagan summarily fired all 11,345 of them. This accentuated a crisis already under way created by an aging inflexible air traffic control system. Part of the solution was to be a system that was more automated, requiring fewer controllers to manage more planes.

Robert Britcher, who worked on the project at IBM Federal Systems, wrote about it in his book, *The Limits of Software: People, Projects, and Perspectives*, where he states that it "may have been the greatest debacle in the history of organized work" (Britcher, 1999, p. 163).

Why did it fail? Cone quotes Pete Marish, a senior analyst at the General Accounting Office:

> It was basically a Big Bang approach, gigantic programs that would revolutionize overnight how [the] FAA did its work. (Cone, 2002)

Cone further quotes Bill Krampf, who worked on the project at IBM Federal Systems:

> We entered [the] software phase without the requirements phase completed. (Cone, 2002)

Krampf's observation, however, is probably a misdiagnosis of the problem. I seriously doubt that completing the requirements phase before starting the work on software would have solved the problem. The idea of completing requirements before engaging in detailed design goes against the grain of one of today's most popular software engineering strategies, called "agile development." In an agile process, requirements are developed along with the software through a series of incremental "sprints," short development efforts with modest partial goals toward the overall project objective. An agile process directly involves the customer and expects requirements to evolve as the project evolves. This way of managing complexity is more realistic than doing specification before design.

Marish's diagnosis is likely more accurate. Wholesale technology replacements generally require too many concurrent paradigm shifts. Indeed, Cone attributes the failure to enormous optimism about emerging immature technology paradigms, including object-oriented design, distributed computing, and the Ada programming language. Cone writes,

> AAS was supposed to be a showcase for Unix-based distributed computing and for development in Ada, a programming language created by the Air Force that became the state-sponsored religion in object-oriented technology, itself a relatively young methodology for writing code in self-contained, reusable chunks. (Cone, 2002)

I find the words "state-sponsored religion" fascinating; they reflect the dogmatic fervor that I frequently encounter in computer scientists who espouse devotion to one or another programming language.

Another dramatic failure of a large engineering project was the U.S. Army's Future Combat Systems (FCS) program. Although there were many reasons that this program failed, one was similar to the reason for the failure of the FAA's AAS program: the program was too ambitious about replacing too many systems all at once. According to a 2012 report by the RAND Corporation,

> Compared to more traditional acquisition strategies, the [systems-of-systems] approach significantly increased both the complexity of the organizations needed to execute the FCS program and the technical

challenges associated with system engineering, software engineering, and system integration. (Pernin et al., 2012)

The FCS program was launched in 2003 with an estimated cost of $92 billion (including the projected cost of a fleet of war-fighting vehicles). By 2009, when Defense Secretary Robert Gates announced that he wanted to scrap the core of the program, the suite of combat vehicles, the estimated cost had ballooned to some $200 billion.

AIS/Net 1000, the FAA's AAS program, and Army's FCS program were all attempting to solve enormously complex problems. When complexity becomes unmanageable, a crisis in the prevailing paradigms becomes apparent. AIS/Net 1000 had as its goal to ameliorate the crisis without inducing a paradigm shift. But that's not how it played out. Instead, the Internet emerged. The other two projects failed in part because they attempted to address crises with a wholesale replacement of many existing paradigms all at once. But technology paradigms grow more organically, more bottom up than top down. They are not imposed on engineers so much as discovered, nurtured, and grown by engineers.

Wholesale simultaneous replacement of several paradigms at once fails because each individual replacement has poor prospects of success. The fact is that most technology innovations fail. We tend to remember only the ones that succeed. It is often impossible to predict which of several competing technologies will eventually prevail, even when it seems clear which technology is more fit.

The layering of paradigms offers a fundamentally creative way to deal with a crisis of complexity. One solution is not to fix a broken paradigm, replacing it with a new one, but rather to build an entirely new paradigm on the scaffolding of the old. We build platforms on top of platforms. Because of separation of concerns, a layer in a paradigm can change, and the effects will only be felt one layer up. In the Internet, for example, at layer 3 of the OSI model, the world is in the midst of migrating from IP version 4 to IP version 6 (no version 5 was ever deployed).

The new version IPv6 changes quite a few fundamental things, including the addresses used to identify nodes in the Internet. IPv4, it turns out, provides only four billion distinct addresses. Given that there are already six billion devices on the Internet, this obviously creates problems. Considerable cleverness is required to reuse addresses without creating ambiguities. IPv6 increases the number of addresses to

$$2^{128} = 340,282,366,920,938,463,463,374,607,431,768,211,456.$$

It simply would not be possible to make such a fundamental change without the separation of concerns offered by the OSI model. This change has no effect, for example, on the JPEG encoding of images transmitted over the Internet.

Similarly, the layers of digital technology in figure 3.3 provide separation of concerns that permits independent simultaneous evolution at all levels. For example, the shift to using FinFETs for transistors had no effect on the design of instruction set architectures, except perhaps to offer opportunities through added capability. I examine this question of opportunity next.

6.3 Crisis and Opportunity

According to Kuhn, observations made in the course of normal science may reveal anomalies, inconsistencies with the prevailing paradigm that governs the normal science. These anomalies can create a crisis that leads to a paradigm shift. In engineering, it is not usually scientific observations that create a crisis. We have already seen that increasing complexity can trigger a crisis. A second trigger is opportunity.

Consider the introduction of the iPhone in 2007. At the time, two of the dominant producers of cell phones were Nokia, a Finnish company, and Research in Motion (RIM), a Canadian company, maker of the Blackberry. These two companies have drastically reduced visibility in the cell phone market today. I've already pointed out that the iPhone introduced no new technology. So why was it such a revolution, summarily overthrowing the old regime?

The "crisis" in the paradigm overthrown by the iPhone was not a crisis of complexity. It was a crisis of opportunity. At the time, cell phones were starting to be used for things other than making phone calls. The Blackberry had captured business markets with its built-in keyboard and email capability. Nokia phones were routinely used, primarily by young people, to send text messages, despite the incredible awkwardness of typing text on a 12-key numeric keypad. At the time, there were even contests for speed texting on such keypads because it required quite a bit of skill.

In 2007, wireless networks had modest ability to carry data, although the emphasis was still on carrying voice signals. That capability was exploited by the Blackberry and by the texting function on other phones, but phones were still primarily for making voice calls. Today, the ability to make voice calls seems like an incidental feature of a smartphone. When I want a voice call with my 20-year-old daughter, I need to exchange several text messages with her to arrange it. She acquiesces only in deference to my age. Her other communications are likely more through Snapchat and other services I've never heard of.

The iPhone came about through a realization of what was possible with the technology of the time. But the real revolution was not replacing the phones of the time with better phones. It was the introduction of a whole new platform, a new layer in the stack of layers of paradigms. Specifically, the real revolution was the introduction of the app development platform. With the introduction of the iPhone, Apple published the specifications that enabled millions of creative programmers around the world to develop applications for the phone and in 2008 launched the App Store to broker the sales of apps to customers.

The nature of a revolution is that its consequences usually cannot be anticipated, but after the revolution its consequences seem inevitable. It is easy to forget today that in 2007, most of us had never heard of apps and app stores, although, as usual with technology innovations, variants of the concept had existed at least since the 1990s. But Apple really made the concept take off. Apple's model has been emulated by every cell phone vendor that still has any significance, down to the details of copying the patented look of the phone, a move that has resulted in an endless string of patent infringement lawsuits and countersuits.

I am quite sure that if we could time travel back to 2007 and assemble the smartest, most creative experts worldwide in a room, they would not be able to anticipate even 10% of the functions that we routinely carry around in our pockets today: instant traffic reports worldwide (go ahead: check the traffic in Budapest right now), airline reservations, banking transactions (even check deposits), tide charts, worldwide weather forecasts, up-to-the-minute mass transit timetables, remote monitoring of our homes, a taxi service, restaurant reviews, a library with millions of books and journals, and many creative games. In addition to all those functions that never before existed, the device replaces several other devices that we previously would have to carry separately, including the phone, the music player, a flashlight, the keys to our house, the video entertainment device (remember the portable DVD player?), a compass, a calculator, an address book, our calendar, a camera, a radio, a notepad, and an alarm clock. Oh yes, and it also sends text messages and email.

The smartphone was not a consequence of a crisis of complexity, it was a consequence of opportunity enabled by millions of transistors on a chip, good digital radios, touch-screen interfaces, and the Internet, all preexisting technologies. The key to the revolution, the decisive battle that won the war, was the app development platform and the app store.

In recent years, we have seen an astonishing number of similarly disruptive revolutions. Amazon put thousands of bookstores out of business and is in the process of threatening the rest of retail. Uber and Lyft have undermined the taxi business. Lulu and other print-on-demand services are threatening the publishing industry. E-books are threatening Lulu and the rest of the printing industry. Libraries are increasingly irrelevant. Travel agencies have almost entirely vanished.

Each of these revolutions entails a paradigm shift. But paradigm shifts do not come easily to people, lending some stability and inertia. Even disruptive changes take some time to play out. According to Statista, an online statistics company, there were still about 28,000 bookstores in the United States in 2012. Although this number is down significantly from more than 38,000 in 2004, it is still a significant number.

Some paradigm shifts replace prior paradigms. Even for taxi services, for example, we are now more likely to summon them using a smartphone app or a web page than via a phone call. But all of these paradigm shifts also build on prior paradigms, leaving them unchanged. Smartphone technology, for example, relies heavily on Internet technology, the latter of which has hardly changed in response to this revolution. To be sure, there are small changes, such as better support at OSI levels 6 and 7 for small screens, but these changes are tiny. The prior paradigm provides a platform for the new paradigm, a situation rarely seen in the scientific paradigm shifts that Kuhn talks about. The transitivity of models makes this possible.

Many examples of failed paradigm shifts also exist. In the 1980s, for example, several university research projects and startup companies were going to disrupt the computer industry with dataflow computers, which presented an entirely different way to define an instruction set architecture (Arvind et al., 1991). These all failed. Perhaps a more curious example is the repeated failure of artificial intelligence (AI) as a field. AI has survived several boom and bust cycles, where unbridled enthusiasm is followed by disillusionment and collapse of investment. Starting in the late 1980s, AI experienced what some researchers in the field called an "AI winter," an allusion to a nuclear winter, and had only fully recovered by about 2010. Perhaps dataflow computers will be similarly resurrected. Such failures fade quickly from our memory (except, of course, for the people most directly involved in the failures).

But failures are a normal and healthy part of intellectual inquiry. The rapid advances of digital technology provide a healthy, thriving ecosystem for mutation, adaptation, and extinction of paradigms. There need not be anything fundamentally wrong with a new paradigm that fails. Unlike scientific paradigms, technology paradigms are not held up to a standard of truth or concurrence with observations of the physical world. Their survival instead depends on many intangibles, including, perhaps most important, the readiness of the public and even the engineers to assimilate the paradigm.

6.4 Models in Crisis

Paradigm shifts in engineering are triggered primarily by crises of complexity and opportunity. These have been relentlessly driven for the last 50 years by the staggering advances in digital technology. They continue to be driven by the increasing interconnectedness of digital devices and penetration of computers into everything we use.

So where are the most pressing crises today? For this question, I can only speculate because I cannot see the future any better than anyone else. But I do see at least two substantial crises looming.

Let me start with a crisis of opportunity. With increasing interconnectedness comes rapidly increasing volumes of data about the state of the world, society, and individuals in society. For example, credit card companies already carefully track most of our purchases, missing only the ones where we pay cash. These companies construct models of our behavior and use those models for various purposes, including to disallow transactions that appear to be anomalous and hence may be fraudulent. For example, if you don't travel much, and a store in China tries to charge a purchase to your credit card, then the transaction is likely to be denied by the credit card company's computers. If you travel a lot, as I do, however, then this same transaction is more likely to be allowed. If you normally buy expensive scotch at boutique stores, then a purchase of Rotgut Moonshine at a store called Payroll Loans & Liquor is also likely to be denied.

These decisions are not made by humans; they are made by computers, the ones that dream (see section 5.6). The computers are running machine-learning algorithms that build models of your behavior by observing your transactions and then classify each subsequent transaction as anomalous or normal based on the probability that the model would generate such behavior.

The credit card example exhibits a contradiction that is common in big data applications. Although we probably appreciate that the credit card company attempts to prevent fraudulent use of our card, many of us find it creepy that the company has built a probabilistic model of our behavior. Similarly, we may like that a map app on our smartphone tells us about nearby restaurants, but we are likely not thrilled to learn that, as a consequence of using the app, Google's computers know where we are.

Now imagine that the computers in your car begin to communicate with the outside world. Some insurance programs already use such communicated information to vary your insurance bill based on usage and style of driving. What if the insurance company sells the information they get from your car to your credit card company? Metcalfe's law is based in part on the observation that aggregated data is more valuable than isolated data. The credit card company may now verify that your car is indeed parked at Payroll Loans & Liquor and either allow the transaction or report your car stolen.

The data being gathered about us by various organizations is growing at a staggering rate. In the United States, privacy laws intended to protect us from misuse of that data are ineffective because these laws have simply resulted in a barrage of small print that every organization is now required to throw at you, knowing that you will not read it. In fact, the U.S. government has exhibited a distinctly double standard, simultaneously trying to strengthen privacy laws and prevent encrypted data communication. Encryption, the government says, interferes with its ability to detect and prevent potential terrorist attacks. Indeed, it no doubt does. Again, we are faced with contradictory requirements.

Many organizations today are collecting but not effectively using vast amounts of data. Consulting and market research company Gartner calls "dark data" the "information assets that organizations collect, process and store in the course of their regular business activity, but generally fail to use for other purposes." The subtext is that those same businesses are missing an opportunity. They should be mining the data. The data has value.

The research and consulting firm Forrester defines "perishable insights" as "urgent business situations (risks and opportunities) that firms can only detect and act on at a moment's notice." Fraud detection for credit cards is just one example of such perishable insights. Once the transaction is allowed, the damage is done. We also saw another example of a perishable insight in chapter 1 in the Wikipedia vandalism detection algorithm, although that one has less privacy cost. More dramatically, dark data in health care and medicine could be much better used to get (literally) perishable insights.

I believe a crisis of opportunity exists today in converting data feeds into insights in time to make effective use of them while either ensuring privacy or at least preserving public trust that the loss of privacy will not be abused. Clearly, this problem is not just technical.

Contradictory requirements demand innovation. Consider that thermostats, door locks, television sets, watches, running shoes, football helmets, books, and, in fact, nearly everything around us is going online. And many devices are acquiring the ability to listen for spoken words and react to those words. And when you read an electronic book, it reads you back. Connecting those devices to the network is likely to deliver real value to us, including, for example, reducing our carbon footprint and vulnerability to terrorists. These potential benefits cannot be ignored. Neither can the risks. The situation is crying for a paradigm shift.

The second crisis I see looming is a crisis of complexity. This crisis is not just about increasing numbers of components but rather about the conjoining of engineered systems that have traditionally used different kinds of models to manage their own complexity.

Consider a modern commercial airplane such as the Airbus A350 or the Boeing 787. These systems are software-intensive with hundreds of microprocessors managing functions that include translating pilot commands into rudder movements, controlling the landing gear, managing cabin pressurization and airflow, managing electric power generation and distribution, and operating the passenger entertainment system. Such an aircraft is a much more complex system than, say, a data center handling Facebook pages. The latter only has to deal with bits and the heat generated by processing those bits. A data center is an information-processing system that can operate almost entirely within the models and paradigms of computer science. But an airplane design conjoins models of aeronautical, mechanical, electrical, and civil engineering, as well as those of computer science. In such a system, the structures of civil engineering interact with the flight dynamics of aeronautical engineering under the control of a software system (computer science) running a feedback control system (electrical engineering). The silos of specialization that are standard in engineering today become an obstacle because the models and paradigms developed in each of these disciplines are incommensurable.

Despite the enormous complexity and challenges of crossing so many silos, Boeing and Airbus both manage to make astonishingly safe and reliable airplanes. How? Today, their design processes and methods are extremely conservative, and regulatory oversight is heavy. Many rules must be followed to get an aircraft certified to carry civilian passengers.

However, these processes reveal some fundamental flaws in the engineering models and methodologies that are available today to aircraft engineers. A symptom of such flaws is a story I first heard from an engineer who had worked on the Boeing 777. This was Boeing's first fly-by-wire airliner, meaning that pilot controls are mediated by a computer. The 777 first entered into service in 1995. According to this engineer, as of the early 1990s, Boeing expected this model of aircraft to be in production for perhaps 50 years, to 2045. The engineer told me that in the early 1990s, Boeing purchased a 50-year supply of the microprocessors for the flight control system so they could use the same microprocessors for the entire production run of the aircraft.

Recall in chapter 4 my observation that hardware is ephemeral. In fact, any particular silicon chip is unlikely to remain in production for more than a few years. It becomes obsolete quickly under the pressure of Moore's law. And when a fab gets updated to leverage a new technology, such as the FinFET, it becomes impossible to produce an identical chip.

But why does Boeing need the chip to be identical? The whole point of the layering of paradigms of figure 3.3 is so that software designs are isolated from changes in the hardware. It should be enough to use any chip that can *correctly* execute the software for the control system.

But the flaw is in the notion of correctness. Starting at the instruction set architecture layer in figure 3.3, and for all layers above that, what it means to *correctly* execute software has nothing to do with how long it takes to do anything. Timing of actions is not part of the programming paradigms used today.

But in a flight control system, the software is directly controlling physical actuators, and in the physical world, timing is important. In fact, in just about every model that an aeronautical, mechanical, or electrical engineer will use, timing of actions is central to the model. The paradigms used by these engineers are incommensurable with the paradigms used by computer scientists.

As a consequence, the layering of figure 3.3 fails to provide adequate abstractions, and hence fails to provide separation of concerns. Boeing is forced to operate without any layering, and hence has to ensure that every airplane carries exactly the same design all the way down to the semiconductor physics layer.

I have subsequently heard similar stories from engineers at Airbus, which has been making fly-by-wire aircraft for longer than Boeing. The engineers at Airbus tell me that they store the microprocessors in liquid nitrogen in an attempt to extend their shelf life by slowing down the natural diffusion processes of the dopants in the silicon.

The complexity of aircraft designs keeps increasing. A key objective in aircraft design is to decrease weight because this reduces fuel consumption, extends range, and improves the carbon footprint. One way to decrease weight is to use more advanced materials in the airframe and more flexible structures. But flexible structures require more tightly coordinated control systems. Timing discrepancies in control systems can create disastrous stresses on airframe structures.

Another way to decrease weight is to reduce the amount of wiring and hydraulic piping. This can be accomplished with more advanced networking, but essentially all of the layers in the OSI model of figure 6.3 above the physical layer also ignore timing, just like the software layers in figure 3.3. As a consequence, aircraft manufacturers are unable to benefit from most of the advances in networking of the last 40 years.

Even with a fixed microprocessor, the fact that timing is irrelevant to correctness from the ISA up means that in software design, aircraft manufacturers also cannot use most of the innovations of computer science from the last 40 years. The FAA prohibits use of object-oriented languages, for example, in safety-critical software. Even interrupts, the standard way that all modern microprocessors get data in and send data out to the outside world, are prohibited. To be sure, interrupts create many subtle software problems. As far back as 1972, Edsger Dijkstra lamented,

> [I]n one or two respects modern machinery is basically more difficult to handle than the old machinery. Firstly, we have got the interrupts, occurring at unpredictable and irreproducible moments; compared with the old sequential machine that pretended to be a fully deterministic automaton, this has been a dramatic change, and many a systems programmer's grey hair bears witness to the fact that we should not talk lightly about the logical problems created by that feature. (Dijkstra, 1972)

Despite this lament, to this day, interrupts remain the primary method for I/O and are central to every modern operating system design. Aircraft manufacturers, therefore, are also excluded from the last 40 years of advances in operating systems.

In view of the failure of most computer science innovations of the last 40 years to address the needs of aircraft designers, I am astonished at their remarkable safety track record. I have enormous respect for the engineers who design these planes. They are stuck with the prototype-and-test style of design that was used by Edison, unable to leverage the transitivity of models, and their prototypes are much more complex than Edison's prototypes.

Figure 6.4 shows a prototype of an Airbus A350, their newest model. Airbus calls this prototype an "iron wing." The prototype includes all parts of an A350 except the airframe, cabin, and engines. This is why it doesn't look like an airplane. It is the guts without the skeleton or skin. The wiring is all exactly the same length as on the real aircraft. The hydraulic tubes are bent as on the real aircraft to get around

Figure 6.4
An Airbus "iron wing" prototype of an A350.

the (missing) airframe structure. When running tests using this prototype, the same generators that are driven by the engines on the real aircraft are driven by artificial engines so that the prototype runs on its own power. This prototype is obviously much more complicated than Edison's lightbulbs, but it is exactly the same sort of concrete prototype.

Aircraft manufacturers are not alone in facing this problem. Modern cars are mostly "drive-by-wire" today, where the driver commands (pushing on the accelerator and brakes and turning the steering wheel) are mediated by a computer before going to the wheels or engine. Automotive designers face the same problems but with far fewer regulatory constraints. Their problems are only going to get worse as automation increases (lane keeping, automated accident prevention, and fully automated driving).

It doesn't stop there. Any modern factory is computer controlled and is similarly a safety-critical system where the timing of actions is essential to safety. Trains are computer controlled. Ventilation, lighting, and fire mitigation systems in buildings are computer controlled. Modern electric power grids, water distribution systems, and communication systems are computer controlled.

In 2006, Helen Gill of the U.S. National Science Foundation coined the term "cyberphysical systems" (CPS) for such systems that combine computing, networking, and physical dynamics. There is clearly today a crisis of complexity for such systems. She launched a major NSF initiative to fund research to address precisely the problems that I have indicated. This program continues, and progress is being made, albeit still primarily in research labs and not yet in industrial production. Gill recognized that conjoining the "cyber" world with the physical world created an enormous crisis of complexity, one that the existing paradigms were poorly equipped to deal with.

What is the origin of the term "cyber" in CPS? The related term "cyberspace" is attributed to William Gibson, who used the term in the novel *Neuromancer*, but the roots of the term CPS are older and deeper. It would be more accurate to view the terms "cyberspace" and "cyberphysical systems" as stemming from the same root, "cybernetics," which was coined by Norbert Wiener (Wiener, 1948), an American mathematician who had a huge impact on the development of control systems theory. During World War II, Wiener pioneered technology for the automatic aiming and firing of anti-aircraft guns. Although the mechanisms he used did not involve digital computers, the principles are similar to those used today in computer-based feedback control systems. His control logic was effectively a computation, albeit one carried out with analog circuits and mechanical parts, and therefore cybernetics is the conjunction of physical processes, computation, and communication. Wiener derived the term from the Greek word for helmsman, governor, pilot, or rudder. The metaphor is apt for control systems.

The term CPS is sometimes confused with "cybersecurity," which concerns the confidentiality, integrity, and availability of data and has no intrinsic connection with physical processes. The term "cybersecurity," therefore, is about the security of cyberspace and is thus only indirectly connected to cybernetics. CPS certainly involves many challenging security and privacy concerns, but these are by no means the only concerns.

My own research at Berkeley includes some examples of NSF-funded research projects under the CPS program. One project that concluded in 2015, called the PRET project (for Performance with Repeatable Timing), designed an instruction set architecture that explicitly includes timing within its programming paradigm. In effect, this project reopened decisions that Fred Brooks made all the way back in the 1960s when he designed the System/360 ISA without any explicit control over timing. The project concluded with a demonstration that it is possible to achieve precise control over timing with no loss of performance and modest cost in hardware. If this architectural approach is adopted in industry, it will enable separation of concerns in the layering of figure 3.3 for cyberphysical systems.

A second example from my research group is a project we called PTIDES, which also concluded in 2015. This project addressed software that is distributed across networks, using the OSI model of figure 6.3, but modifying the paradigms to explicitly control timing. But I will spare you the details.

Despite progress in the research labs, the crisis of complexity remains for cyberphysical systems, and the crisis of opportunity remains for data science. In the remaining chapters, I examine just how far we can push the layering of models. All the techniques we know of today have limitations, and understanding these limitations is essential to a complete understanding of technology revolutions. These limitations hint at opportunities for innovation.

II YIN

7 Information

··· in which I examine the concept of information, what it is and how to measure it; and in which I introduce Claude Shannon's way of measuring information and show that information cannot always be represented digitally.

7.1 Pessimism Becomes Optimism

In chapter 2, I emphasized the importance of keeping distinctly separate in our minds the model and the thing being modeled. Unfortunately, this is really hard to do. Because so much of our thought process is structured around models, we have an enormous backdrop of unknown knowns. But a failure to make this separation inevitably leads us to invalid conclusions.

Engineers choose their models and then seek physical realizations that are faithful to those models. For this task, we need models that we can understand. Although we have developed, over centuries, a huge arsenal of models and ways of building models, I will show that the number of possible models in this arsenal is tiny compared with the number of models possible in theory. There is no end to the possible engineering innovations.

A scientist, in contrast to an engineer, tries to find or invent a model (which can take the form of a "law of nature") to match a given physical object or process. A timeless goal in science has been to find a small number of such models that can somehow "explain" everything in the universe. In some sense, whereas an engineer strives to grow the number of relevant models (those for which we can build a faithful physical realization), the scientist tries to shrink the number of relevant models (those needed to explain the natural world).

Unfortunately for science, this timeless goal is unachievable. We already know that nature is capable of creating processes that are at least as sophisticated as software on computers because humans and our computers, after all, exist in the natural world. In chapter 8, I will review Alan Turing's classic result that it is impossible, in general, to tell what a program will do merely by looking at the program. This finding alone crushes the optimism that any small set of rules can explain everything in the universe because it shows that we have no way to explain the behaviors of some programs that exist in the universe. All we have are the programs themselves.

The situation for science is exacerbated by the fact that mathematical models, which are not exactly the same as computational models and form the bedrock of scientific explanations of the natural world, are incomplete in a similar manner to software, as I will explain in chapter 9. Kurt Gödel's classic incompleteness theorems show that any system of mathematical models that is potentially rich enough to explain the natural world will be either inconsistent or incomplete. "Inconsistent" means that it has statements that can be shown to be both true and false. "Incomplete" means that it has statements that cannot be shown to be true and cannot be shown to be false.

The goal of an engineer is not to explain the natural world but rather to create artifacts and processes that have never before existed in that natural world. Engineers need only to explain the systems they design, not the systems given to them by nature (at least not as their primary task). They build physical systems to match their models rather than the other way around. If the modeling toolkit is rich and expressive enough, and if the space of physical systems is vastly larger than the modeling toolkit, then there is plenty of room for innovation.

Engineering, of course, requires science. As we try to synthesize new physical realizations from our models, we will learn more about nature as we try to understand how the physical realizations deviate from the predictions of the model. Engineering, therefore, can provide a set of guideposts for science by exposing phenomena that nature has not happened to expose.

The transistor is a good example of this reversal, where engineering drives science. I know of no natural occurrence of a transistor that has ever been found. But the engineering effort to create an electrically controllable switch has led to a much deeper scientific understanding of how electricity behaves in materials. That scientific understanding has in turn enabled better engineering models, bootstrapping a process of progress that cannot occur by just passively observing systems that nature happens to give us. The limitations exposed by Turing and Gödel do not impede this progress. Instead, they simply assert that this process will never be complete. There will always be room for more progress.

So far in this book, I have argued that digital technology and computation provide a rich medium for such creative work. But like the scientist, the engineer is subject to fundamental limitations of both computation and mathematical models. For engineers, these limitations do not undermine any timeless mission. Engineers have the luxury that they can try to *avoid* systems they cannot explain. Software engineers, for example, usually try to write programs that will exhibit behaviors they can explain, avoiding programs that Turing showed are inexplicable. But to avoid them, they need to understand the limitations of their modeling toolkit. Thus, in contrast to the previous chapters in this book, in the next few chapters I will explain what software and mathematical models *cannot* do.

I focus on digital technology and computation, which are fundamentally about processing information. But what is information? Only with a clear notion of information can we understand what can and cannot be done with digital technology. Hence, information is the subject of the rest of this chapter.

7.2 Information-Processing Machines

A computer program is a model. Ultimately, it models electrons sloshing around in silicon and metal. But as I've pointed out, there are so many levels of abstraction between software as a model and semiconductor physics that viewing software as a model of electrons sloshing is not useful. In fact, software is more usefully viewed like mathematics. It is a formal model, existing in its own self-scaffolded world. Like mathematics, it is a powerful model. We can do a lot with software.

But we can't do everything. In fact, I will show you in chapter 8 that despite the incredible power of software, we can do almost nothing with it, in the sense that no matter how much we do with it, there are vastly many more things we *cannot* do with it.

Many futurists and technology enthusiasts exaggerate the capabilities of software. To pick on one, in his book *Tools for Thought*, Howard Rheingold states,

> The digital computer is based on a theoretical discovery known as "the universal machine," which is not actually a tangible device but a mathematical description of a machine capable of simulating the actions of any other machine. Once you have created a general-purpose machine that can imitate any other machine, the future development of the tool depends only on what tasks you can think to do with it. (Rheingold, 2000, p. 15)

Rheingold draws the same conclusion that I do, that technological progress is limited by humans rather than by technology but for the wrong reason. Rheingold actually misrepresents the history of computing. There is no universal machine, mathematical or otherwise. What he is actually referring to is known as the "universal *Turing* machine," which is capable of simulating any *Turing* machine, not any machine. There are machines that are not Turing machines, like my dishwasher, for example.

But wait, I'm sure that Rheingold would object now that my dishwasher is not an *information-processing* machine, and his book is about *information-processing* machines. Dirty dishes are not a form of information (or maybe they are; see section 8.4 in the next chapter). So what is an information-processing machine?

The first question we have to answer, the one I focus on in the rest of this chapter, is, what is information? Merriam-Webster has several definitions, but to me the most relevant to software is:

> 2. b: the attribute inherent in and communicated by one of two or more alternative sequences or arrangements of something (as nucleotides in DNA or binary digits in a computer program) that produce specific effects.

The key in this definition is "one of two or more alternative sequences or arrangements." Information is the resolution of alternatives. When there are two alternatives, for example, a transistor can be on or off or a coin can yield heads or tails, "information" is the determination of one of the alternatives.

With a little thought, I hope you can see that this is consistent with an intuitive notion of information. If, for example, I don't know whether my colleague Fred is married to Sue, then there are two "alternative arrangements." If you tell me that Fred is married to Sue, then I have received information from you in the sense that what you have conveyed to me resolves these alternatives. Notice that it is still information, even if you lied to me. You have conveyed a resolution to the alternatives, and whether that resolution is true is a separate issue.

Merriam-Webster also gives the following definition:

> 2. d: a quantitative measure of the content of information; specifically: a numerical quantity that measures the uncertainty in the outcome of an experiment to be performed.

Measuring information is a relatively recent development, usually credited to Claude Shannon. In 1948, while working at the storied Bell Labs, Shannon published a paper called "A Mathematical Theory of Communication" in the *Bell System Technical Journal*. This paper launched the field of information theory (Shannon, 1948). In this paper, Shannon used probability theory (about which I will say more in chapter 11) to come up with a measure of the amount of information contained in a sequence of bits and the amount of information that can be conveyed over an imperfect communication channel. Solomon Golomb, who was hugely influential in subsequent development of coding and information theory, and to whom I owe the "drilling through the map" metaphor, remarked that Shannon's influence cannot be overstated: "It's like saying how much influence the inventor of the alphabet has had on literature" (Horgan, 1992).

Suppose, by the first definition from Merriam-Webster, that there are exactly two "alternative arrangements" for something. Fred is either married or not. A coin toss can yield heads or tails. Then once we have resolved these alternatives, we have received one bit of information, exactly representable by one binary digit, 0 or 1. Can information always be represented by bits?

Note that the physical world rarely gives us exactly two alternative arrangements for anything. A coin toss could result in the coin falling into a pond, sinking to the bottom, and embedding in the mud in a vertical position, with neither heads nor tails being up. Even in Fred's case, Fred may be married to Sue, in that there are papers filed with the local courthouse, but living with Joe and wishing that his state allowed him to be married to Joe. Marriage is a model of a social structure, and only the model can have a binary choice between exactly two arrangements. The physical world is messier. We need to be careful to keep the map (the legal institution of marriage) distinct from the territory (Fred's actual situation).

Nevertheless, Shannon measured information in bits. As it turns out, this measure only works well when the alternative arrangements are discrete and finite, or when attempting to communicate over an imperfect channel. Suppose that instead of telling me whether Fred is married (one bit of information), you tell me the temperature in the room right now. How many bits of information have you conveyed? This question is impossible to answer without making many more assumptions. How much do I already know about the temperature in the room? Is the number of possible temperatures finite? Perhaps I only care about the temperature within a degree or so. Then the number of possible messages you will convey to me is certainly finite. But have you really conveyed the temperature? Does the temperature itself in the room entail information? Could it have an infinite number of possible values?

These are all difficult questions. Even when the number of alternative arrangements is finite, the amount of information conveyed by resolving the alternatives is not always obvious. Shannon noticed that the amount of information conveyed depends not only on the number of alternatives but also on the likelihood of the alternatives. I will next explain how Shannon measured information in units of bits when the number of alternatives is finite.

7.3 Measuring Information

Suppose that we have an unfair coin that almost always comes up heads. Then observing a head does not convey much information. Most outcomes are heads, so you will not be surprised to see heads. Suppose that we toss the coin 20 times and get the following outcomes:

HH HT HH HH HH TH HH HH HH HH

where "H" represents heads and "T" represents tails. We can code this sequence of outcomes using binary digits 0 and 1 as follows:

HH	HT	HH	HH	HH	TH	HH	HH	HH	HH
11	10	11	11	11	01	11	11	11	11

This encodes the outcomes using 20 bits. It is a quite literal encoding, using a 1 to represent H and a 0 to represent T. However, because tails is much less likely than heads, there are relatively few tails, so Shannon noticed that this sequence can be encoded with fewer than 20 bits by using a less literal encoding. For example, suppose that we group the coin tosses in pairs, as above, and encode pairs of results according to the following table:

HH	0
TH	10
HT	110
TT	111

In other words, when we get two heads in a row, we will represent that fact with a single bit, 0, rather than two bits, 11. If we get TT, then we will represent that with three ones in a row, 111. These codes are carefully chosen so that any sequence of bits can be unambiguously decoded into a sequence of coin tosses. For example, 010 represents HHTH, four outcomes.

The previous sequence of coin tosses can now be represented as follows:

HH	HT	HH	HH	HH	TH	HH	HH	HH	HH
0	110	0	0	0	10	0	0	0	0

The more likely "HH" pairs are encoded efficiently with just one bit, whereas the less likely sequences require more bits. This encoding requires only 13 bits rather than the direct encoding, which requires 20. Shannon noticed that if heads are much more likely than tails, then most of the time this alternative encoding will require fewer bits than the direct encoding. So the amount of information in 20 unfair coin tosses is usually less than 20 bits, Shannon observed.

Shannon also noticed that ordinary English-language text could also be encoded more efficiently. A great deal of redundant information is found in a sequence of letters and spaces. If I text you a message saying, "i lv u," I'm pretty sure you will understand it.

Note that during World War II, Shannon worked on coding schemes for secret communications, including codes used by Roosevelt and Churchill for trans-Atlantic conferences. His cryptography work no doubt lay the groundwork for information theory because it made it clear to Shannon that a message could be encoded in many ways. Some ways would result in a more efficient encoding (fewer bits), and some would be difficult to read if you didn't know the code. If you don't know the code given in the earlier table, then the sequence 0110000100000 is hard to interpret as HHHTHHHHHHHTHHHHHHHHH. For most of us, it is hard even if we *do* know the code, unlike "i lv u," which requires no explicit listing of the code.

Our encoding may not *always* work well, however. For example, suppose we get a sequence of 20 tails in a row. The prior encoding will require 30 bits instead of the 20 that the direct encoding requires because each pair TT will be encoded by three bits, 111. This is unlikely, but it is still possible. Using probability, which I will talk about in chapter 11, we can estimate how unlikely this is. If the coin tosses are all independent (they do not influence one another), and on average 1 in 10 tosses comes up tails, then the probability of 20 tails in a row is 10^{-20}. In chapter 11, I will discuss what this really means, but for this example it simply means that if you repeat the experiment of tossing 20 coins 10^{20} times (100 quintillion times), on average, you can expect one occurrence of 20 tails in a row. This outcome is very rare indeed. In fact, we can also use probability to show that most outcomes will require fewer than 20 bits. But I will spare you that nerd storm.

Based on earlier work by Hartley (1928), Shannon used this observation to come up with a quantitative measure of the amount of information conveyed by observing a single coin toss. According to Shannon, if our unfair coin comes up heads, then when we observe this fact, we get $-\log_2(0.9) \approx 0.15$ bits of information. Here, 0.9 is the probability of heads, indicating that 9 out of 10 tosses yield heads, on average. Because heads are much more likely than tails, we can take this as a measure of our surprise or what we have learned, or, in short, information. When we observe heads, we get 0.15 bits of information, much less than one bit. We are not surprised. The information in observing an outcome of tails is $-\log_2(0.1) \approx 3.32$ bits, where 0.1 is the probability of tails. We are much more surprised when we see tails. So seeing tails conveys more information, 3.32 bits, than seeing heads, 0.15 bits.

If instead the coin were fair, then the probability of T would be 0.5, meaning that, on average, half of all coin tosses yield T. The Shannon information in observing T is therefore $-\log_2(0.5) = 1$ bit of information. For a fair coin, every toss gives us one bit of information. It is more surprising than seeing H for the unfair coin and less surprising than seeing T for the unfair coin.

Why the logarithm? This seems kind of arbitrary, a rabbit pulled out of a hat. But a logarithm has a nice property, which is that for any two numbers a and b, $\log_2(ab) = \log_2(a) + \log_2(b)$. Logarithms turn multiplication into addition. Consider a pair of unfair coin tosses that turn out to be TH, tails followed by heads. How much information is in that result? Well, we get 3.32 bits from the T and 0.15 bits from the H, so the pair presumably conveys the sum or 3.47 bits of information. When you receive a sequence of unrelated messages, the information conveyed is the sum of the information in each of the messages (we assume each coin toss has no effect on the outcome of the next coin toss).

Suppose that we toss two coins simultaneously and observe TH. What is the information content in that? To apply Shannon's theory, we need to determine the probability of TH. If the coin tosses are independent (one does not affect the other), then the probability of TH is the product of the probabilities for T and H, or $0.1 \times 0.9 = 0.09$. This probability is slightly less than 0.1, indicating that slightly fewer than 1 in 10 times tossing two coins we will see TH. So the Shannon information conveyed by observing TH is $-\log_2(0.09) \approx 3.47$. Because a logarithm turns a product into a sum, this is the same as the sum of the information we get from each coin toss, $-\log_2(0.1 \times 0.9) = -\log_2(0.1) - \log_2(0.9)$. This is why Shannon used a logarithm. It makes the information content in two identical coins tossed simultaneously the same as the information content in two sequential tosses of the same coin.

The logarithm base 2 was used by Shannon so that the information measure would be in units of bits. If you use the natural logarithm instead, then the information measure has units of "nats." If you use base 10, then it has units of decimal digits. In all cases, however, it measures information.

You might ask why there is the annoying minus sign everywhere. There are two reasons. One is that probabilities are always less than one,[1] and the logarithm of a number less than one is negative. We would prefer a positive number to quantify information, and the minus sign gives us that. The second and more important reason is that the quantity of information should increase as the event gets more rare. Without the minus sign, it would decrease, and the relationship between information and rarity would be backward.

If on average 1 in 10 coin tosses yields tails, then Shannon said that the average information in a single coin toss is

$$-0.9 \log_2(0.9) - 0.1 \log_2(0.1) \approx 0.47 \text{ bits.} \tag{64}$$

Equation (64) is just the average of these two information quantities, weighted by their probability, and hence it is the average information in one coin toss. This means that each coin toss conveys about 0.47 bits of information rather than one bit of information, on average. In theory, therefore, we may be able to come up with an encoding that will represent 20 coin tosses with only $20 \times 0.47 = 9.38$ bits, on average. Shannon showed, in fact, that we cannot do any better, so 9.38 bits per 20 coin tosses is the limit. No encoding scheme will do better than this, on average, so the average amount of information in 20 unfair coin tosses is 9.38 bits.

Equation (64) represents what Shannon called the "entropy" in a coin toss. Shannon chose the term "entropy" for this because the mathematical structure of his formula resembles the formula that had previously been used for a concept called "entropy" in thermodynamics. In a profile of Shannon, Horgan writes:

> The great mathematician and computer theoretician John von Neumann persuaded Shannon to use the word entropy. The fact that no one knows what entropy really is, von Neumann argued, would give Shannon an edge in debates over information theory. (Horgan, 1992)

Thermodynamics studies the macroscopic properties of materials (especially gasses) in terms of the microscopic properties (especially molecules in motion). The notion of entropy originated in 1870 with the work of physicist Ludwig Boltzmann in Austria, James Clerk Maxwell in Scotland, and Josiah Willard Gibbs in the United States. Entropy is the central concept in the second law of thermodynamics. That law asserts that the entropy in the universe (or in any isolated system within the universe) tends to increase. To Boltzmann, Maxwell, and Gibbs, entropy was a measure of randomness or disorder in a physical system. Physical systems tend

---

[1] Probabilities are less than one because we can say "one out of ten coin tosses comes up tails" (probability $1/10 = 0.1$), but we would not say "eleven out of ten coin tosses comes up tails" (probability $11/10 = 1.1$).

inexorably toward greater randomness, where eventually all outcomes are equally likely.

Specifically, Boltzmann and his contemporaries modeled the degree of randomness (entropy) in a macroscopic system (e.g., a volume of gas) as

$$S = k \log(M), \qquad (32)$$

where M is the number of possible states of the microscopic system (a collection of gas molecules) that are consistent with the observed macroscopic properties of the system (like the temperature and pressure of the gas). The constant k is a scaling constant called the Boltzmann constant. If each of the M states is equally likely, then the probability of each state is $1/M$, and this becomes $-k \log(1/M)$, which is similar to Shannon's entropy.

At least two significant differences can be found between Boltzmann's entropy and Shannon's. One is the constant multiplier k, which simply changes the units with which we are expressing entropy. Shannon used bits as his units,[2] whereas Boltzmann used joules per degree kelvin (energy per temperature). Because temperature is actually energy, Boltzmann's entropy is dimensionless. Dimensionless quantities do not generally measure something in the physical world, but they can be useful when making comparisons. Boltzmann's entropy allows us to *compare* the entropy in two scenarios, and the second law of thermodynamics is all about comparing entropies. Entropy at one point in time is higher than entropy at an earlier time. The law assigns little meaning to the actual numbers, only to their relative magnitudes. An exception is that when the entropy is zero, there is a distinct physical meaning. For the entropy to be zero, we need $M = 1$, which means that there is only one possible state. For an ideal gas, this occurs exactly at a temperature called absolute zero, approximately $-460°$ F or $-273°$ C, where all motion stops.

A second difference between Boltzmann's entropy and Shannon's concerns the notion of the "number of possible states." In Shannon's model, this notion is well defined because the very notion of the number of possible states is part of the model. When considering a coin toss, Shannon did not consider the unlikely possibility that the coin would land on its edge in the mud, yielding neither heads nor tails. Instead, the "coin toss" is just a physical metaphor for a *model* where there are exactly two possible outcomes. The notion of probabilities for these outcomes, a notion I consider in more depth in chapter 11, is also part of the model and hence is well defined.

In Boltzmann's case, however, what is the "number of possible states" of a physical gas? This is well defined at absolute zero but more difficult to pin down at achievable temperatures. Boltzmann assumed each molecule in the gas had a physical position

---

2 In fact, the standard term for Shannon's units is shannons, in his honor, but many people continue to use bits.

and velocity and the state of the molecule was captured by these numbers. How many possible values are there for the position and velocity of a molecule? In Boltzmann's time, there was no physics that would bound the number of possibilities for these numbers to a finite set. The more recent development of quantum mechanics changes the situation. At least for a closed system with well-defined boundary conditions, quantum mechanics does yield a finite number of states. But for all but the tiniest systems, the number of states is enormous. Moreover, precisely defining the boundary conditions is treacherous and risks confusing the map with the territory. Despite these considerable subtleties, many people have associated Boltzmann's entropy with Shannon's quite closely and concluded that the world is digital. I will examine this question in the next chapter.

But sticking to Shannon's self-contained and well-defined notion of entropy, which exists entirely in the world of models, entropy is a good measure of the uncertainty, randomness, or disorder in a system. If the coin is fair (heads and tails are equally likely), then the entropy is 1 bit, so no encoding scheme can do better than the direct encoding, on average. This is the highest level of uncertainty, randomness, or disorder that a coin-toss system can have. At the other extreme, if the coin is extremely unfair, and you only ever get heads, then the entropy is zero. No information at all is conveyed by observing a coin toss. Certainty is high, there is no randomness, and the coin-toss system is perfectly ordered. The result of a coin toss can be encoded with no bits at all because we already know the outcome.

Shannon's quantification of information content and his choice to express this quantification in bits had an enormous impact on communication theory, computer science, and even philosophy. But it is easy to forget that the theory I've described here applies much more readily to scenarios where the alternative arrangements are finite and distinct. What happens if the alternative arrangements offer an infinite number of possibilities? For example, what if the position of each molecule in Boltzmann's gas can be any point in a volume of space? This set of alternative arrangements is not finite. A direct adaptation of Shannon's entropy to scenarios with a continuous range of possible outcomes has to be interpreted more carefully. I do that in the next section.

I apologize in advance that the next section is a bit more technical. The short story, should you wish to skip to the next chapter, is that information cannot always be represented as binary data. Hence, there is a notion of information that is out of reach for computers.

7.4 Continuous Information

Equation (64) gives the entropy of a random experiment (an unfair coin toss) that has exactly two possible outcomes, one with probability 0.1 and one with probability 0.9. Shannon showed that this entropy can be interpreted as the minimum number of bits required to encode an outcome of the experiment, on average. Equation (64) states that roughly half a bit (0.47 bits) is required to encode each outcome of the unfair coin toss. Equivalently, on average, each bit in a sequence of bits can encode the results of roughly two coin tosses. This requires clever encoding of a sequence of outcomes of the experiment, but with such encoding, it quantifies the amount of information gleaned from observing each coin toss, about half a bit, on average.

A fair coin, in contrast, has an entropy with value 1, so on average one bit is needed to encode each outcome. In this case, no clever coding is needed because we can just encode heads with 1 and tails with 0. Every coin toss yields one bit of information.

Formula (64) is a sum of two quantities, each of the form $-p\log_2(p)$, where p is the probability of one of the two outcomes, and the negative of the logarithm quantifies the amount of information in that outcome. The more rare the outcome, the more information it carries. It is easy to generalize this idea to a random experiment with more than two possible outcomes, such as the toss of a pair of dice. The sum in (64) will simply have one term of the form $-p\log_2(p)$ for each possible outcome with probability p.

But what if we have a random experiment that can have an infinite number of possible outcomes? Suppose, for example, that some variable is equally likely to have any real-numbered value between $-a$ and a for some positive real number a. What is the entropy of this random experiment, and how many bits are required to encode an outcome?

The formula for entropy is easy to adapt, where the summation of terms of the form $-p\log_2(p)$ in equation (64) is replaced with an integration over a continuous range of possible values. Specifically, the entropy of a continuous random experiment is given by the formula

$$H(X) = -\int_\Omega f(x)\log_2(f(x))dx. \tag{16}$$

$H(X)$ represents the entropy of a random experiment that we name X. Bear with me.

The form of equation (16) is similar to equation (64). An integral, after all, is just a sum over a continuum of an infinite number of values (I will return to the idea of a continuum in chapter 9). The integration is a sum over the set Ω of all possible values x that the experiment might yield. Each term being summed by the integral has a form similar to that of the terms in equation (64), $-p\log_2(p)$, except that probabilities p have been replaced with *probability densities* $f(x)$. The term $f(x)$ is the probability density at x, where x is one of the possible outcomes of the experiment. A probability density, just like a probability, reveals the relative likelihood that the experiment will yield certain outcomes. I will talk about probability densities more carefully in chapter 11, but loosely, if for two possible outcomes x and y we have that $f(x) > f(y)$, then the experiment is more likely to yield an outcome *in the vicinity of* x than in the vicinity of y. The phrase "in the vicinity" reflects that this is a probability *density* not a probability.

The continuous entropy of equation (16), like the discrete entropy of equation (64), represents the average amount of information obtained by observing outcomes of the experiment. Like discrete entropy, outcomes that are more rare (values of x where $f(x)$ is lower) carry more information than values that are more likely. *It is no longer correct, however, to interpret this entropy as specifying the average number of bits required to encode an outcome of the experiment.* In fact, *every* outcome will require an infinite number of bits to encode. An outcome of a continuous random value is not representable exactly with binary numbers, unlike a discrete random value.

Consider a simple example, a random experiment that can yield any real number between $-a$ and a. Assume that every value in that range is equally likely. For this experiment, the probability density f is plotted in figure 7.1 for the particular case where $a = 4$. The plot shows that outside the range between $-a$ and a, $f(x) = 0$, indicating that those values have zero probability, whereas inside that range, $f(x) = 1/8$, indicating that all values inside the range are equally likely.

The probability density function indicates the relative likelihoods of outcomes of the experiment. An area under the plot, like the shaded area in the figure, indicates the probability that the experiment will yield an outcome inside a range. The area of the shaded rectangle in the figure is $1 \times 1/8 = 1/8$, which indicates that the probability of an outcome between 1 and 2 is one eighth. This indicates that one in eight outcomes will lie in this range.

The total area under the plot for any probability density f is required to add up to one because that total area indicates the probability of *any* outcome, and the experiment must yield some outcome. In the figure, the total area under the plot for f is a rectangle with width 8 and height $1/8$, so as required, the area under f is $8 \times 1/8 = 1$.

For the probability density function of figure 7.1, the entropy $H(X)$ in equation (16) is easy to calculate. An integral is just finding the area under a curve, and the "curve" in this case is not curvy. It is a rectangle. Without boring you with the details,

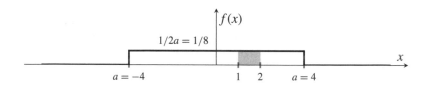

Figure 7.1
Probability density function for a uniform continuous random experiment.

with the uniform probability density of figure 7.1, the entropy becomes

$$H(X) = -\log_2(1/2a) = \log_2(2a) = 3. \tag{8}$$

If we were to erroneously interpret this as the number of bits required to encode an outcome of X, then we would conclude that three bits are sufficient. But this should worry us. How could three bits distinguish between an infinite number of possible outcomes?

A discrete entropy as in equation (64) is always zero or positive. It cannot be negative. A probability p is always between zero and one, and the logarithm of a number between zero and one is always negative, so $-p\log_2(p)$ is always nonnegative. A sum of nonnegative numbers is always nonnegative.

For the continuous random experiment, the situation is a bit different. The entropy $H(X)$ can be positive or negative. Suppose, for example, that X has a probability density function similar to figure 7.1, but instead of $a = 4$, it has $a = 1/4$. Then when x is in the range $-1/4$ to $1/4$, the density must be $f(x) = 2$. It has to be 2 because the total area under f must be 1. But now you can verify from equation (8) that $H(X) = -\log_2(2) = -1$. The entropy is negative! This further underscores that continuous entropy does not represent the number of bits required to encode an outcome. How could we encode the outcome of an experiment using a negative number of bits?

When a continuous entropy is negative, this should not be interpreted as meaning that negative information is conveyed by an observation of the experiment. In fact, infinite information (in bits) is conveyed. It should instead be interpreted to mean that an experiment with negative entropy conveys less information than one with positive entropy. Indeed, when $a = 4$, there are more possible values for x than when $a = 1/4$, so a (perfect) observation in the first scenario conveys more information than a (perfect) observation in the second. But neither bundle of information can be encoded using bits.

There is an interesting special case when an experiment has only one possible outcome, for example, a coin toss that always yields heads because both sides of the coin are heads. In the discrete case, the entropy is zero. The sum in equation (64) will have only one term, $-p\log_2(p)$, where $p = 1$. But the logarithm of 1 is 0, so an experiment with only one possible outcome has entropy equal to zero. It requires zero bits to encode because we already know the answer. This makes sense.

What if we have a continuous experiment where there happens to be only exactly one possible outcome? In other words, the experiment is not actually random, like our coin with two heads. Suppose, for example, we have a continuous experiment that happens to always yield $x = 0$. To model this, we can use the probability density function of figure 7.1 and let a become arbitrarily close to zero. As a gets small, the height $1/2a$ of $f(x)$ in the range $-a$ to a gets large to ensure that the area under f remains 1. As a approaches zero, $f(x)$ approaches infinity for $-a \leq x \leq a$. As a consequence, as a approaches zero, $H(X) = -\log_2(1/2a)$ approaches minus infinity.

So for a continuous random experiment, an entropy of *minus infinity* indicates that no information is conveyed by a measurement. This is different from a discrete random experiment, where an entropy of *zero* indicates that no information is conveyed. The difference between zero and minus infinity is huge so confusing the two forms of entropy will yield drastically erroneous conclusions. If we insist on trying to compare these two forms of entropy, then we need to acknowledge that there is an infinite offset between them. It takes infinitely more bits to encode the continuous outcome than the discrete one.

What does it mean when the continuous entropy is zero? Not much. It just means that there is more information than if the entropy had been negative and less information than if the entropy had been positive. You can verify that if $a = 1/2$ in figure 7.1, then $H(X) = 0$.

Why does each outcome of the continuous experiment require an infinite number of bits to encode? I will fully address this subtle question in the next chapter. The short answer is that there are vastly more possible outcomes than there are finite bit sequences. There are just not enough finite bit sequences to assign a unique bit sequence to each possible outcome. This will become clear in the next chapter, but for now I ask you to take my word for it so that I can explain a truly remarkable insight, due to Shannon.

In the same 1948 paper, Shannon observed the rather obvious fact that any *noisy* observation of an experiment yields less information than a perfect observation. "Noise" is a term that engineers use for extraneous factors that creep into measurements so that the measurements are imperfect. Noise is unavoidable in any measurement of the physical world.

But Shannon's truly remarkable observation was that a noisy observation of a continuous-valued experiment yields *much less* information, and that the information it yields *can* be represented with a finite number of bits. Although outcomes of the experiment contain information that requires an infinite number of bits to represent, any noisy observation only reveals a finite number of bits. Thus, all *measurements* of the physical world can be encoded with binary digits, assuming all measurements are noisy.

But this does not imply that the physical world can be encoded with binary digits. That would be confusing the map for the territory. It *may* be true that a physical system can be encoded with bits, although I personally doubt it (see section 8.4 on digital physics in the next chapter), but it is not finite entropy that makes it true. Continuous entropy is finite even though its continuous variable *cannot* be encoded with a finite number of bits.

Shannon's result that a noisy measurement reveals a finite number of bits of information is known as the "channel capacity theorem." Shannon was considering communication problems, where a quantity known at one point in space is to be conveyed via an imperfect communication channel to another point in space. One of his central results is that any channel that adds noise can only convey a finite amount of information, measured in bits, for each observation of the output of the channel. The output of the channel is a noisy observation of the input to the channel.[3]

So how much information is conveyed by a noisy measurement? Consider the experiment X with probability density function as shown in figure 7.1 and entropy $H(X)$ as given in equation (16). Let Y represent a noisy *measurement* of X. Let x represent a particular outcome of experiment X and y represent a particular measurement of that outcome. Because the measurement is noisy, it is likely that y is close to x but not exactly equal to x. The measurement y tells us *something* about an outcome x but not everything. So how much does it tell us?

If we have a model for the measurement noise, then given some specific measurement y, we can come up with a probability density function that represents the relative likelihoods of values x that could have yielded the measurement y. This new probability density is called a "conditional probability density" because it is a valid probability density for x only once we have a measurement y.

Suppose we know that our measurement apparatus adds noise no bigger than $1/2$. This means that given a measurement y, it must be true that x is within the range $y - 1/2$ to $y + 1/2$. Suppose further that we have the measurement $y = 1.5$ right in the middle of the grey region in figure 7.1. We can conclude that the actual value of x is equally likely to be anywhere in the grey region, in the range from 1 to 2. It cannot be anywhere else because the measurement apparatus does not add noise larger than $1/2$.

Armed with this knowledge, once we observe $y = 1.5$, the actual value of x is still random (it is not known), but now it is equally likely to be anywhere in the grey region. We have gained information because without the measurement, it was equally likely to be anywhere from -4 to 4. Now we know that it is equally likely to be in the range from 1 to 2. We have significantly narrowed the range.

How much information have we gained? Intuitively, the grey region is one eighth of the total possible region for x. Hence, our measurement tells us that the actual value for x is in one of eight possible equally sized regions. We can distinguish eight regions using three bits because three bits have eight distinguishable patterns: 000, 001, 010, 011, 100, 101, 110, and 111. So it seems we have gained three bits of information. Shannon shows us that we actually gain slightly more than three bits of information, on average, because measurements that are close to the edge of the region, near -4 or 4, will yield more information than measurements in the middle of the region under this noise model. For example, if our measurement happens to be $y = 4.5$, then the *only* possible value for x is 4, so with this (extremely unlikely)

---

3 Any text on digital communication will cover this topic of channel capacity, including one that I coauthored (Barry et al., 2004, p. 123).

measurement, we have gained a huge amount of information. We have achieved certainty.

How did Shannon determine the information conveyed by a noisy measurement? Once a measurement is taken, we have a new conditional probability density function for X. Figure 7.2 shows the conditional probability density given a measurement $y = 1.5$ and noise limited to $\pm 1/2$. We can use that new density in equation (16) to calculate the entropy. Let's call this entropy $H(X|Y)$, which we read as "the entropy remaining in X given a measurement Y." Shannon's channel capacity theorem then tells us that the information yielded by a measurement is, on average,

$$C = H(X) - H(X|Y). \tag{4}$$

$H(X)$ is the information we would gain with a perfect observation, and $H(X|Y)$ is the information that is *not* revealed by the experiment. In other words, $H(X|Y)$ is the remaining randomness after the measurement. The truly astonishing thing about this theorem is that for a wide range of models of measurement noise, the difference $H(X) - H(X|Y)$ represents information that *can* be encoded with a finite number of bits, even if the original outcome x of experiment X cannot be encoded with a finite number of bits. The information revealed by the experiment, in bits, is finite, although the information in the actual system, in bits, is infinite. In forming the difference $H(X) - H(X|Y)$, both quantities have an infinite offset compared with discrete entropy, but the offsets cancel, and the difference becomes a discrete entropy. This insight is truly remarkable.

For our particular example, where $a = 4$, we have determined that $H(X) = 3$. Calculating $H(X|Y)$ precisely is a bit tedious, so I will spare you the details, but our intuition holds up, and $H(X|Y)$ turns out to be slightly less than 0. Hence, C in equation (4) turns out to be slightly larger than 3, indicating that our measurement reveals slightly more than 3 bits of information on average.

It is now worth considering some special cases. Suppose the measurement is perfect. In this case, $H(X|Y)$ is minus infinity because once a measurement is taken, there is no remaining randomness in X. Hence, C is infinite regardless of the value

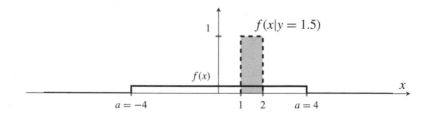

Figure 7.2
Conditional probability density function (dashed line) given a measurement $y = 1.5$.

Chapter 7

of $H(X)$ (as long as $H(X)$ is not also minus infinity). As a consequence, *a perfect observation of a continuous random experiment yields an infinite number of bits of information.*

Suppose that our experimental apparatus is hopeless, and a measurement yields *no* information about X. In this case, $H(X|Y) = H(X)$ because the randomness after observation is the same as before. Hence, $C = 0$. A hopelessly bad measurement yields zero bits of information about X.

It is also easy to see that $H(X) - H(X|Y)$ cannot be negative because the randomness (the uncertainty) $H(X|Y)$ after measurement cannot be more than the randomness (uncertainty) $H(X)$ before measurement. Hence, making a measurement never reveals a negative number of bits of information.

In short, an outcome of a continuous random experiment has information, but that information cannot be encoded in a finite number of bits. A noisy observation of the outcome of continuous random experiment, however, *can* be encoded with a finite number of bits. That number is given by the Shannon channel capacity theorem, equation (4). So the question arises whether the physical world presents scenarios where variables can have values in a continuous range. There is real risk here of confusing the map and the territory, so I defer this question to a more careful analysis in the next chapter.

As I argued in chapter 2, in an engineering use of models, we seek physical realizations that match a model. Models that represent information digitally are extremely useful, and thanks to the digital technology outlined in chapters 4 and 5, we know how to make physical systems that are faithful to this digital representation of information. In the scientific use of models, in contrast, we seek models that match the physical world. In this use, the assumption that all information is digital and can be represented in bits is questionable (see section 8.4 in the next chapter). This assumption is demonstrably untrue if continuous quantities exist in nature.

In chapter 11, I will examine the meaning of probability, which underlies Shannon's notion of information. Fundamentally, probability is a measure of uncertainty, the lack of information. The entropy in a system quantifies exactly how much information we lack about the system. Put another way, entropy quantifies how much information can potentially be gained by observing the system. But there are two distinct and incomparable measures, discrete and continuous entropy. Only discrete entropy measures information in bits.

In the next chapter, we will look at machines whose sole purpose is to process digital information. I will argue that even if we restrict our attention to the digital world, leaving out my dishwasher, software is still limited. It cannot perform most information-processing functions.

8 The Limits of Software

··· in which I explain what software cannot do and show that the number of
information-processing functions is vastly larger than the number of possible computer
programs; and in which I explain the Church-Turing thesis, which shows that there
are useful information-processing functions that are not realizable by software. But it
does not follow that if a function is not realizable by software, then it is not realizable by
any machine. Here, I am forced to confront the paradigm of "digital physics," which
argues that the physical world itself is somehow software or equivalent to software.

8.1 Universal Machines?

Computers are information-processing machines. The previous chapter studied
what information is and how to measure it. I showed that information is not
necessarily representable digitally, at least in theory. In practice, the inevitability
of measurement noise and the possibility that the physical world is digital may lead
to the conclusion that information in the physical world *can* always be represented
digitally. If we go a step further and assume that all transformations of information
in the physical world are performed in essentially the same way as in computers,
then it is impossible in principle to do more than what can be done by software.
This conclusion, if it were true, would be quite remarkable because it turns out
that software can do little compared with what we can imagine is possible. In this
chapter, I will explain the limits of software and why I believe that we *can* (and do)
do things that software cannot.

The set of all computer programs, each of which is a model, is actually a tiny set.
The size of that set is the same as the *smallest* of all the infinite sets that Georg Cantor
identified in the late 1800s. Cantor, a Russian-German mathematician, showed that
some infinite sets are vastly larger than other infinite sets.

It is difficult to reason about size when talking about infinity. In fact, Cantor spent
12 years attempting to prove that all infinite sets have the same size (Smullyan, 1992,
p. 219). He failed! In the process, he developed a remarkable insight that I will use to
show how much smaller the set of all computer programs is compared with the set of
functions that we might be interested in implementing on computers. Consequently,
although we can do an extraordinary amount with software, it's nothing compared
with what is possible if we do not limit ourselves to this smallest of infinite sets.

There is, however, a potential caveat that I am forced to confront. Since the
development of information theory and the theory of computing, a branch of thought
has emerged that some people call "digital physics." Digital physics postulates that
nature does not and cannot have a continuous range of possibilities. Some of the

stronger forms of digital physics postulate that going beyond what software can do in principle is physically impossible. In practice, however, going beyond what software can do is clearly possible today and in the foreseeable future. Software cannot realize my dishwasher, for example. Nevertheless, I will conclude the chapter with a discussion of digital physics.

For now, let's put aside the question of whether the physical world is digital and just consider information-processing functions where the inputs are binary numbers and the outputs are binary numbers. In fact, let's consider an even smaller set of functions, those whose input is a finite binary number and whose output is just a *single* zero or one rather than a sequence of zeros and ones. Such functions are called "decision functions" because for each particular input, say 010101, the function will say either YES (1) or NO (0). The function makes a decision.

In the 1930s, the young English computer scientist Alan Turing defined the set of "effectively computable" functions to be those decision functions that can be computed algorithmically (in a step-by-step fashion) using a machine that is now called a Turing machine.[1] In principle, a Turing machine is realizable by a modern computer that has a sufficient amount of memory. Independently of Turing, in 1936, Alonzo Church, an American mathematician, came up with a different model than the Turing machine that yields exactly the same set of effectively computable functions. The fact that two different models result in the same set of effectively computable functions suggests that there is something special about this particular set of functions.

There are many possible Turing machines, each of which may compute a different decision function. In Turing's formulation, each such machine can be encoded by a finite sequence of bits, much the way machine code encodes a computer program as a finite sequence of bits (see chapter 5). For example, the sequence 000111 might represent a Turing machine that computes a particular decision function. Turing showed that there is a "universal Turing machine," a Turing machine that can implement any other Turing machine. For example, if 000111 encodes a Turing machine, and we would like to know what decision that machine makes for the input 010101, then we can concatenate the bits specifying the machine code and the input to get 000111010101. Providing that combined bit pattern as the input to a universal Turing machine yields the answer that the machine 000111 would have given. So the bit pattern 000111 encodes the program, and the universal Turing machine is the computer that executes the program. So a "universal" Turing machine is simply

---

1 Later, like Claude Shannon, Alan Turing worked on cryptography during World War II. Turing played a central role in intercepting German communications that were encrypted using a machine called the Enigma. Turing led a troubled life, including being prosecuted for homosexual acts in 1952, which were illegal in the United Kingdom at the time. In 1954, he took his own life at age 41. In his few short years, however, he transformed the landscape of computing. The highest honor in computer science, the Turing Award, is named after him.

a programmable Turing machine where the program can encode any other Turing machine.

The effectively computable functions constitute the set of decision functions that can be realized by a universal Turing machine. What is now called the Church-Turing thesis states that any function that a human can compute using a systematic, step-by-step process, given enough pencil and paper and enough time, is one of the effectively computable functions.

Given enough memory and time, any modern computer can also compute any effectively computable function. So any modern computer is a universal Turing machine, except that it may run out of memory. But memory has become so cheap and plentiful that this caveat carries little weight.

The question remains whether a modern computer can do *more* than compute effectively computable functions. Many people, including Rheingold quoted in chapter 7, mistake the Turing-Church thesis to state that it cannot. In fact, Rheingold goes further to state that *no* machine can do more than compute effectively computable functions. Turing and Church considered only machines that operate on digital data, and only machines that compute algorithmically, via a step-by-step process. We routinely build machines that satisfy neither of these properties, such as my dishwasher.

A universal Turing machine implements *algorithms*, step-by-step processes, where each step changes the state of the machine discretely. The word "algorithm" comes from the name of the Persian mathematician, astronomer, and geographer, Muhammad ibn Musa al-Khwarizmi (780—850), who was instrumental in the spread of the arabic system of numerals that we all use today. An algorithm is a step-by-step calculation procedure, a recipe. The notion of an algorithm is central to computer science, but it is important to recognize that an algorithm is a *model* of what a machine does. In a modern computer, what is really happening is electrons sloshing around.

The notion of a step, a discrete operation that takes a calculation toward its conclusion, is an abstraction. Most processes in the physical world do not proceed in a sequence of discrete steps.[2] Even a human walking, from which we get the concept of "steps," is not actually discrete because each step evolves as continuous motions that begin with leaning forward and lifting a leg. But the digital machine abstraction considered in chapter 4 abstracts the underlying continuous physical processes in semiconductors as a discrete sequence of steps. An algorithm is an abstraction that ignores the messy continuous-world details of the computer. In this abstraction, a step does not take time, and there is no notion of being halfway through a step. A step occurs *atomically*, meaning indivisibly and instantaneously.

---

2 If you accept a strong form of digital physics (see section 8.4), then every process in the physical world does, in fact, proceed in a sequence of discrete steps. But for most purposes, at the macroscale at which we interact with the physical world, this does not provide a useful model of physical processes.

A second important feature of an algorithm is that it reaches a conclusion. That is, it halts, giving a final answer. The purpose for an algorithm is to determine that answer. An algorithm that realizes a decision function must halt, giving the answer 0 or 1. If it does not halt, then it does not realize the decision function.

To compute an effectively computable function, the program executing on a universal Turing machine must halt to deliver the final answer. Modern computers routinely run programs that are not designed to halt, such as the operating system (see section 5.4). These programs do occasionally halt, but we describe such halting as a "crash," making it quite clear that this was not intended. A program that does not halt does not realize an effectively computable function. Nevertheless, even an operating system is a composition of Turing computations, chunks of computation that are each algorithmic, digital, and halting.

An operating system implements *interactive* behavior, which is quite different from implementing a decision function. Peter Wegner, a computer science professor at Brown University, has written extensively arguing that interactive programs can do more than algorithms (Wegner, 1997). An interactive program does not have access to all of its input when it starts executing. Input may be provided by the program's environment while the program is running, and the program can provide outputs to the environment before it halts (if it halts at all). Hence, the program becomes able to probe the environment, providing stimulus to the environment, watching its reaction (which will be provided as input), and adapting its own behavior accordingly. Turing's model included no such interaction. In his model, the input is unaffected by the output, so no dialogue with the environment is considered in the model. But still, every interactive program is a composition of algorithmic, digital, and halting chunks.

Nevertheless, if an interactive program is interacting with the physical world (i.e., the program is part of a cyberphysical system; see chapter 6), then the timing of the actions of the program will affect the overall behavior of the system. Such a program is clearly *not* a Turing computation because Turing's model includes no notion of time. The timing of the program's actions must be considered part of its "output" because the timing affects the behavior of the system. But Turing's model includes no notion of time, so the behavior is not expressible within his model. For such interactive programs, Wegner was clearly correct that they are not algorithmic.

An interactive program may be interacting with another interactive program. These two programs may even be executing on the same machine if the machine is capable of multitasking,[3] as most modern computers are. Such a pair of programs

---

3 Multitasking means that the computer executes several programs at once rather than completing one program before executing the next one. Without multitasking, a computer could only ever execute at most one nonhalting program. The word "multitasking" has even spread into the vernacular to refer to humans simultaneously handling more than one task.

is said to be *concurrent*. Again, the timing of actions may affect the overall behavior, so Turing's model requires some extension to include such systems.

Robin Milner, who appeared in chapter 5 as the author of the ML programming language, in his Turing Award lecture in 1975, observed that concurrent programs cannot be modeled simply as functions from inputs to outputs, as (halting) Turing machines can be. Their chunks can be modeled as such functions but not their overall behavior.

So Wegner and Milner argue that modern computers can do things that a universal Turing machine cannot do, at least when a program is viewed holistically. Their arguments continue to be debated, but even if you accept them, modern computers, even if you endow them with unbounded memory, are *still* not universal machines. They still cannot do many things. In fact, I will show that there are *vastly* more things that a computer *cannot* do than things a computer *can* do. The reason is quite simple: the number of possible computer programs is much smaller than the number of things we might want to do. This is true even if we limit ourselves to implementing decision functions and much more obviously true if we consider functions where timing matters. Even in the limited case of decision functions, there are vastly more decision functions that neither a computer nor a Turing machine can compute than decision functions they can compute. Decision functions that cannot be realized by software on a computer are said to be "undecidable."

8.2 Undecidability

Recall that a decision function takes as input a finite binary integer, such as 010101, and produces a binary result, 0 or 1. I can prove to you that no computer can realize all decision functions even if the computer has unbounded memory. Because a modern computer, given enough memory, can do anything that a Turing machine can do, Turing's universal machines are also unable to realize all decision functions.

In fact, *almost all* decision functions are undecidable, or, equivalently, almost all decision functions are not effectively computable. But it is a logic error to conclude from this that *no* machine can realize these functions or other functions beyond decision functions. To draw that conclusion requires accepting a strong form of digital physics.

I can give you a rather simple proof. First, let's assume that we have a modern computer with an unbounded amount of memory. This hypothetical computer can implement any function that a universal Turing machine can implement.

You might already be objecting. Unbounded memory? Any computer will eventually run out of memory, so actually it will not be able to do everything that a universal Turing machine can do. But even if the computer *does* have unbounded memory, it is *still* not a universal machine. Specifically, this hypothetical computer cannot realize all decision functions. To show this, I will use a variant of a clever argument that Cantor used that is called "diagonalization."

To show that not all decision functions can be implemented by our unbounded-memory computer, I just have to find one decision function that is not implemented by any program for that computer. In section 8.3, we will see that there are many more than one decision function that cannot be implemented by any program, but we only need one to show that our unbounded-memory computer is not a universal machine.

First, note that I can create a list of all the possible inputs to our decision functions:

0
1
00
01
10
11
000
001
. . .

Each input is a finite sequence of bits. This list will get very long. Infinite in fact. But hopefully you can see that every possible input, every possible finite bit sequence, will be somewhere on this list.

Every program for any modern computer is represented as a sequence of zeros and ones. As a consequence, every program will also be somewhere on this list. Not all elements of the list are valid programs, but every valid program must be on the list.

Some of these valid programs produce as output just one number, 0 or 1, for each possible input. Let's call those programs "decision programs." Every decision program is on the list. Clearly, every decision program realizes a decision function, but not every decision function has such a program.

Let's call the first decision program on the list P_1, the second P_2, and so on. We see now that we can assign a name of the form P_n for every decision program, where n is an integer. Each of these programs yields a decision for each of the inputs in the previous list. For example, it might be that program P_1 yields the following outputs:

$P_1(0) = 0$
$P_1(1) = 0$
$P_1(00) = 1$
$P_1(01) = 1$
. . .

I can now find a decision function that is not implemented by any decision program. This is a contrarian function, so I will call it C. The decision function C yields the following decisions for each input:

$C(0) = \neg P_1(0)$
$C(1) = \neg P_2(1)$
$C(00) = \neg P_3(00)$
$C(01) = \neg P_4(01)$
$\dots$

where the symbol $\neg$ is logical negation. It converts a 0 to a 1 and vice versa, like the NOT gate in chapter 4. So if $P_1(0) = 0$, then $\neg P_1(0) = 1$.

This function is contrarian because for each possible input, it yields the opposite of what one of the decision programs yields. Notice now that C is different from every decision program. It is different from P_1 because its output differs from P_1 for input 0, it is different from P_2 because its output differs from P_2 for input 1, and so on. So the decision function C is not implemented by any computer program. Hence, not all decision functions can be implemented by our infinite memory computer. Our proof is concluded.

Notice that C *resembles* the functions that a computer *can* realize, in that its input is a sequence of bits and its output is one bit. It seems that a computer should be able to realize C, but I have just shown that it can't.

The "effectively computable" functions are exactly those decision functions that are realized by one of the decision programs P_n in the list. It is quite a remarkable fact that for any modern computer, regardless of its particular machine code structure, the set of effectively computable functions is essentially the same. Any computer that can realize all the effectively computable functions is said to be *Turing complete*. Any modern computer, if we extend it to have unbounded memory, is Turing complete. The decision function C is not an effectively computable function, or, equivalently, C is *undecidable*. There is no computer program that can make the decisions that C makes.

You might be protesting that C is not a useful decision function. It's just a cute corner case, an academic exercise. I have two rebuttals. First, I will show in the next section that there are vastly more functions like C than there are effectively computable functions. Second, many *useful* decision functions are known to be undecidable. Turing gave one: Turing's useful function takes as input a binary encoding for a Turing machine concatenated with an input string and returns 0 if the Turing machine halts and 1 if it does not. A Turing machine that does not halt just keeps executing forever without ever giving a final answer. Turing's useful function solves the so-called *halting problem*, telling us whether a program will halt for a particular input. This is clearly something that could be useful to know. Turing showed that this function is undecidable.

It is surprisingly easy to show that the halting problem is undecidable. If you will again indulge me a brief nerd storm, the proof is another diagonalization argument, as shown earlier. First, assume we have a computer program called H that solves the halting problem. If H were to exist, then it would be given two inputs: a binary encoding B of a program and a binary encoding I of an input to that program. All that H has to do is return 0 if B halts for input I and return 1 otherwise. It must return these values after a finite number of steps. We can't wait forever for H or its answer would be useless.

Because B and I can be concatenated into a single input bit sequence, H is a decision program. Let's write the concatenated input BI, and let's write $H(BI)$ to mean "execute program H on input BI." For H to actually solve the halting problem, it must itself halt for any input. Given any input BI (or at least any input BI where B is a valid program), it must return 0 or 1.

Because B and I are both sequences of bits, if H exists, then we should certainly be able to evaluate $H(BB)$, and it should return 0 if B halts with input B, and 1 otherwise. Now recall that every program is encoded as a bit sequence, and every bit sequence is in the list of bit sequences 0, 1, 00, 01, 10, 11, 000, $\cdots$. The list must contain *every* program that, given input BB for any valid program B, halts and returns 0 or 1. Let's call the first such program on the list F_1, the second one F_2, and so on. We can now show that H must be different from every one of these F_n programs, and hence H cannot be a program that halts and returns 0 or 1 for every input BB. It cannot solve the halting problem.

For each F_n on the list, we construct a new program T_n, a rather annoying program like the contrarian program C shown earlier. It uses F_n as a subprogram, but it does not itself implement an effectively computable function. It fails to halt for some inputs. Specifically, given a valid program B, the first thing T_n does is evaluate $F_n(BB)$. By assumption, $F_n(BB)$ always halts and returns 0 or 1. If $F_n(BB)$ returns 1, then $T_n(P)$ returns 0. Otherwise, and here is where T_n gets annoying, T_n loops forever and never returns anything. It fails to halt. In pseudocode,[4] T_n looks like this:

```
if (Fn(BB) == 1) {
    return 0
} else {
    loop forever
}
```

We can now show that none of the F_n in the list solves the halting problem or, equivalently, that the function that F_n implements is different from the function that

4 Programmers use "pseudocode" to refer to a sketch of a program that is not written in any particular programming language. Its purpose is to communicate intent to other humans.

H implements for every $n = 1, 2, \cdots$. To do this, suppose we evaluate the annoying function T_n on its own binary encoding. That is, we evaluate $T_n(T_n)$. Does this halt? If it does, then $H(T_nT_n)$ should return 0; otherwise it should return 1. What does $F_n(T_nT_n)$ return? Well, we don't know because F_n is just some arbitrary program realizing an effectively computable function. But we do know that $F_n(T_nT_n)$ halts and returns something.

Suppose that $F_n(T_nT_n)$ returns 1. Then T_n halts and returns 0, so $H(T_nT_n) = 0$. So in this case, F_n is not the same as H. They return different values. Suppose instead that $F_n(T_nT_n)$ returns 0. Then T_n annoyingly does not halt, so $H(T_nT_n) = 1$. So again F_n is not the same as H.

Because H is different from F_n for all n, H is not on the list of programs that return 0 or 1 for any input BB. Hence, H cannot be a program that solves the halting problem. Whew. Survived another nerd storm.

A direct consequence of the undecidability of the halting problem is that in any programming language, there will be valid programs where we can't tell what the program will do just by looking at the program. We cannot tell whether the program will ever halt and give an answer.

The philosophical consequences are profound. Programs and their executions on computers exist in the physical world. So Turing has shown us that there are physical processes that we cannot fully "explain." A full explanation of some physical process should, it seems, tells us *why* it exhibits some behavior. But Turing showed us there are processes for which we cannot even tell *whether* they will exhibit some behavior, much less why.

A fundamental theme of this book is that we must insist on keeping separate in our minds the map and the territory. Any "explanation" of the physical world or of what a computer does is a map. It is a model. A computer program is a model of its own execution. Turing showed us that this model can never fully explain the thing it models. A program does not fully explain its own execution because you cannot tell what it will do just by looking at the program. The world is what it is, and every model, every explanation, every map, and every program we construct is a human invention distinct from the thing that it models. The set of all maps is incomplete, in that there are properties of the physical world that cannot be mapped.

Nothing that I have said should be construed to undermine the value of models or maps. It leaves open the question of whether programs, as models, can *describe* all processes, even if we now know that they cannot *explain* all processes. The question remains whether our hypothetical unbounded-memory computer is a universal machine. If you *define* "computation" to mean exactly those *decision functions* that are *effectively computable*, then our unbounded-memory computer can realize all "computations." But this argument is circular. We have defined "universal" to mean "it does everything that it does." This is why the Church-Turing thesis is called a *thesis* and not a theorem. It simply states that the effectively computable functions are the ones we can compute with an (idealized) computer. Nothing about this thesis says that there is *no* machine that can realize *C*. All we have shown is that a *computer*, as defined by the digital technology of chapters 4 and 5, cannot realize more than a small subset of decision functions.

It turns out to be a surprisingly controversial topic whether there could be a machine to realize functions like *C*. A whole community has emerged that looks at what some people call "hypercomputation" or "super-Turing computation" that specifies (mostly hypothetical) machines that compute functions that are not effectively computable. For example, Blum et al. (1989) describe a hypothetical machine that is similar to an ordinary computer except that it operates on real numbers rather than digital data. But its execution is still algorithmic, so many of the same questions and answers carry over from Turing machines, such as the question of whether a program ever halts.

Martin Davis, an American mathematician whose PhD thesis advisor was Alonzo Church, vocally debunks the very idea of hypercomputation, except possibly as a pure theory exercise (Davis, 2006). Even his arguments, however, are set squarely in the framework of algorithmic operations on finite bit sequences. He ignores timing, for example, presumably assuming that timing is irrelevant, and he assumes that input and output are discrete. He observes, for example, of any machine that produces a non-Turing-computable infinite sequence of natural numbers (which can be encoded as finite bit sequences), "no matter how long this goes on, we will see only a finite number of these outputs" (Davis, 2006). This assumes the outputs are rendered to the observer as a list (which is discrete) of natural numbers (also discrete). Is this the only way to present a result of a computation? It is the way computers present results, but what about other machines? My dishwasher does not present its results as a list of numbers.

Ultimately, I believe that any conclusion that no machine can realize functions like *C* is an act of faith, a position I will defend more strongly in the next chapter. Like many acts of faith, this one requires ignoring evidence against it. Clearly, computers are not universal machines because they can't do what my dishwasher does. Surely my dishwasher is a machine, albeit not an *information-processing* machine. In fact, it is a cyberphysical system because it includes a computer working in concert with mechanical and hydraulic systems. A dishwasher that presents clean dishes as a list of numbers won't sell well. But actually, even the resistors and inductors considered in chapter 2, as modeled by Ohm's and Faraday's laws, are not realizable by computers because they do not operate on binary data and do not operate algorithmically.

Nevertheless, many authors, not just Rheingold and Davis, subscribe to this faith of universal computation. In fact, it's a powerful religion these days, with many believers. Turing and many others since have conjectured that even the human brain, and hence human cognition, is realizable by a universal Turing machine. As I will show in the next chapter, I cannot prove this conjecture false, but I believe it is extraordinarily unlikely. This faith of universal computation can only be true if a strong form of digital physics is true, and even then it will not lead to useful models.

Turing himself actually described a hypothetical machine that *can* realize C. He called it an "oracle machine" and assumed it would not be implementable. I can explain a simple version of it. Assume infinite memory. Technically, this assumption is stronger than the assumption of *unbounded* memory that we made for the universal Turing machine because "unbounded" means that we have all the memory we need, whereas "infinite" means that we can actually store an infinite list of bits. Even so, this new machine will prove not to be a Turing machine.

Suppose that the infinite memory initially contains a table of all the outputs that C produces for each input. This memory is like an oracle, all knowing. That is, the first entry in the table has one bit with value $\neg P_1(0)$, which gives us $C(0)$; the second entry has value $\neg P_2(1)$, which gives us $C(1)$; and so on. Now, given any input, such as 010101, the machine simply goes down the table until it finds the entry matching this input and produces the corresponding output $C(010101)$. For any input, this procedure can be accomplished in a finite number of steps, so this machine will always produce an output eventually. It realizes the decision function C. In fact, just by providing a different initial table, this hypothetical machine can realize all decision functions, so it's much more "universal" than a universal Turing machine.

I have already pointed out, but it bears repeating, that Turing never intended the word "universal" in "universal Turing machine" to mean that these machines could do everything. Turing's machine is universal in the sense that its program, which defines the function it computes, is part of the input to the machine. A nonuniversal Turing machine, by contrast, is not programmable. It computes exactly one function, and that function is built into it. So Turing's "universal" means simply that the machine can be programmed to compute any effectively computable function. It is a mistake to reinterpret "universal" to mean that it can do everything.

In what sense is this oracle machine *not* a Turing machine? The key issue is the memory containing the table. Turing did not include in his description of a universal Turing machine the ability to initialize an infinite memory. Modern computers also do not have this capability. But does this mean that no machine can do this?

To conclude that there cannot be such a machine, I would have to assume that it is physically impossible to construct something that "remembers" any infinite sequence of bits. Is it? Only by accepting digital physics can I conclude that this is impossible.

Suppose I want to remember the number π. A binary encoding requires an infinite number of bits. Suppose that I can cut a steel rod so that its length is exactly π

meters. In 1799, a platinum bar was placed in the National Archives in Paris and became, for many years, the standard definition of one meter of length. So it is not so far fetched to use the length of a rod for memorizing a value. Haven't I just made a memory that stores an infinite number of bits of information?

Of course, I will run into a number of practical physical problems with this memory. The length of the rod will vary with temperature (and with passing gravitational waves, apparently), and I would need a clear and precise specification of what "one meter" is to interpret the length of the bar as "π meters." Measuring the length precisely will be difficult or even impossible due to quantum mechanical uncertainty laws. And if there is any noise at all in the measurement process, then Shannon's channel capacity theorem from the previous chapter, equation (4), shows that the information *conveyed* by a measurement contains only a finite number of bits.

But just because I can't measure the length doesn't mean that the rod doesn't have a length. Moreover, even if the length of the rod changes over time, as it does so, if it progresses fluidly from one length to another, then the length at each instant will have some dependence on the lengths at prior instants. Isn't such dependence a form of memory? If the length changes fluidly from one value to another, then at least some of the intermediate lengths would require an infinite number of bits to be represented precisely under any units of length, be they meters, inches, furlongs, or any arbitrary unit.

The form of the information is not in bits. Where are the bits? But then again, where are the bits in a computer? A key premise in the concept of information is that the form in which it is stored is not important. This is why information can be conveyed and copied. The idea of conveying and copying information depends on the assumption that the recipient has the same information as the originator, although the physical form of the information is obviously distinct in the recipient. Modern computer memories store bits electrically or magnetically, using the techniques of chapter 4 to abstract the messy physics into digital models. When information is conveyed from one computer to another, the form of storage can change, for example, from electrical charges to magnetic polarization. When designing computer memories, engineers have no need to assume that the underlying physics is discrete. So what is wrong with the means by which my rod stores the number π? Nothing. If I insist that the form be as a list of bits, then I've already assumed the conclusion, that such a memory is not possible.

An astute reader can use many remaining problems to challenge my position. Suppose, for example, that you require that in order for something to be deemed to be "information," it must be possible to convey or copy it. Then we run into a fundamental problem. Shannon's channel capacity theorem, equation (4), tells us that if there is *any* noise in the channel over which the information is conveyed or copied, then only a finite number of bits of information gets through. With this observation, you could define "information" to only include things that can be represented with a finite number of bits. This requires rejecting the use of continuous entropy (section 7.4) as a measure of information.

I am a teacher. I know quite a lot about a few things. I think what I know is "information." But I also know that I routinely fail to convey this information to my students. Some of it gets through, but not all of it. I am privileged to work with extremely smart students, and many of them creatively misunderstand what I am trying to convey and come up with insights that I never had. And sometimes they fail to convey those insights to me. But this does not make what I know and what they know any less "information." It is still information even if it is not conveyed.

So I believe there is plenty of room for doubt that a universal Turing machine is a universal information-processing machine. This is probably a minority opinion today, and I will try to defend it better in the next section. The root of my doubt lies in the mathematical notion of cardinality of infinite sets. The fact is that the set of all computer programs is a small infinite set. In fact, the size of this set is equal to the size of the smallest infinite sets that mathematicians know about. Much bigger infinite sets exist. To assume that all the machines we can make are limited to this smallest of infinite sets, I have to assume digital physics. To assume that all machines that *nature* can make (or has made) are also so limited, I have to reject the existence of *anything* continuous in nature. This requires accepting one of the stronger forms of digital physics. I will now try to explain just how unlikely this limitation is by examining the notion of cardinality. I will then directly confront the idea of digital physics in section 8.4. In chapter 11, I will explain why it is that when some hypothesis is unlikely to be true, we must demand much stronger evidence before accepting the hypothesis than if the hypothesis is a priori likely to be true.

8.3 Cardinality

A mathematician uses the term "cardinality" for the size of a set.[5] A set with two items, for example, has cardinality two. This rather trivial concept becomes interesting only when we consider sets that have an infinite number of items, such as the set of all computer programs.

In the previous section, I showed that there is at least one decision function that is not implementable by any computer program. Using Cantor's results, we can show more strongly that an infinite number of decision functions are not realizable by any computer program. Even more strongly, we can show that *vastly more* decision functions *cannot* be realized than decision functions that *can* be realized by a computer program.

---

5 I recommend the wonderfully readable book on this subject by Raymond Smullyan called *Satan, Cantor & Infinity* (Smullyan, 1992). I first learned from this book that Cantor was trying to show that all infinite sets have the same size when he discovered that they did not.

This result depends on Cantor's observation that not all infinite sets have the same size. To put this in intuitive terms, consider the set of all nonnegative integers, $\mathbb{N} = \{0, 1, 2, 3, \cdots\}$. The symbol $\mathbb{N}$ is shorthand for this entire infinite set of integers. There are clearly a lot of them, an infinite number, in fact, as indicated by the ellipsis "$\cdots$," which can be read "and so on." This set is called the set of "natural numbers" presumably because someone thought that negative and fractional numbers were somehow unnatural.

Consider now the set of all "real numbers," commonly given the symbol $\mathbb{R}$.[6] This set includes all the elements of $\mathbb{N}$ but also many more numbers. It includes negative numbers, fractions, and irrational numbers (numbers such as π that cannot be represented using fractions). Clearly $\mathbb{R}$ is a bigger set than $\mathbb{N}$. But how much bigger?

First, we need to be clear on what we mean by the size of an infinite set. In Cantor's notion of the sizes of infinite sets, two infinite sets A and B are said to have the same size if we can define a one-to-one correspondence between the elements of the sets. A one-to-one correspondence means that for every element of A, we can assign a unique element of B to be its partner. For example, consider the set $\mathbb{N}$ and another set $\mathbb{M} = \{-1, -2, -3, \cdots\}$. We can establish a one-to-one correspondence as follows:

$$\mathbb{N}: \quad 0 \quad 1 \quad 2 \quad 3 \quad \cdots$$
$$\updownarrow \quad \updownarrow \quad \updownarrow \quad \updownarrow \quad \cdots$$
$$\mathbb{M}: -1 \; -2 \; -3 \; -4 \; \cdots$$

Each and every element of one set has a unique partner in the other set. So Cantor's observation is that we can declare these two sets to have the same size.

Interestingly, we can also establish a one-to-one correspondence between the set $\mathbb{N}$ and a subset of itself containing only even integers, $\mathbb{E} = \{0, 2, 4, 6, \cdots\}$:

$$\mathbb{N}: 0 \; 1 \; 2 \; 3 \; \cdots$$
$$\updownarrow \updownarrow \updownarrow \updownarrow \; \cdots$$
$$\mathbb{E}: 0 \; 2 \; 4 \; 6 \; \cdots$$

Every element of both sets is represented, so these two sets also have the same size, although the second set omits half the elements of the first. This oddity is a property of *infinite* sets. They have the same size as many of their own subsets.

---

6 Although the concept of real numbers is quite old, appearing in the ancient Greek work of Archimedes and Eudoxus, who was a student of Plato's, the modern formalization of the concept is relatively recent, dating to the nineteenth-century work of Weierstrass and Dedekind. It is actually quite a subtle concept. According to Penrose (1989), "to the ancient Greeks, and to Eudoxos in particular, 'real' numbers were things to be extracted from the geometry of physical space. Now we prefer to think of the real numbers as logically more primitive than geometry."

Cantor denoted the size of the set $\mathbb{N}$ by the symbol $\aleph_0$, where $\aleph$ is the first letter of the Hebrew alphabet, aleph. The subscript 0 indicates that this is the size of the smallest known infinite sets. Mathematicians pronounce $\aleph_0$ "aleph null."

Interestingly, many sets have size $\aleph_0$, including the natural numbers $\mathbb{N}$, the integers, and even the rational numbers. The set of binary sequences listed in the previous section, let's call it $\mathbb{B} = \{0, 1, 00, 01, 10, 11, 000, \cdots\}$, also has size $\aleph_0$. Hopefully, you can see how to establish a one-to-one correspondence between this set and $\mathbb{N}$.

A set with size $\aleph_0$ is called a "countably infinite set" because there is a one-to-one correspondence with the set of counting numbers $\{1, 2, 3, 4, \cdots\}$. Hence, we can "count" the elements of the set, although we will eventually tire of doing so because of its infinite size.

For any set with size $\aleph_0$, every infinite subset of that set also has size $\aleph_0$. Consequently, the size of the set of all computer programs is also $\aleph_0$. This is because every computer program is in the set $\mathbb{B}$, and an infinite number of such programs is possible.

Now things get really interesting. Cantor, to whom I owe the diagonalization arguments in the previous section, showed that there are infinite sets with vastly bigger size than $\aleph_0$. In particular, Cantor showed that the set $\mathbb{R}$ of real numbers has no one-to-one correspondence with $\mathbb{N}$. He used a diagonal argument similar to what I used in the previous section. Mathematicians say that the set $\mathbb{R}$ is "uncountable." Moreover, there are even bigger infinite sets, such as the set of all functions that map real numbers into real numbers.

There are many uncountable sets besides $\mathbb{R}$. In fact, the set of decision functions from the previous section is also uncountable. It has the same size as $\mathbb{R}$. To show this, we can establish a one-to-one correspondence between the set of real numbers and the set of decision functions. We could do that here, but I'll spare you that nerd storm and ask you to take my word for it.

As with $\mathbb{N}$, a proper subset of $\mathbb{R}$ may have the same size as $\mathbb{R}$. For example, the size of the set of real numbers between zero and one is the same.

An uncountable set is strictly larger than a set with size $\aleph_0$. In light of this result, it is not surprising that not all decision functions can be realized by computer programs, even though decision functions only involve binary digits. The set of decision functions is uncountable, and the set of computer programs is countable and therefore much smaller. If more decision functions than the decidable ones are realizable, then computers are not universal information-processing machines.

Now notice that any machine that deals with real numbers will also not be realizable by computer programs. Hence, to consider computers to be universal information-processing machines, we have to exclude real numbers from our notion of "information." This rather drastic step goes against almost all tradition in mathematics, science, and engineering. We should not accept this step lightly.

Recall from the previous chapter that information can be measured in bits if the number of alternative arrangements being distinguished is finite. That is, information in bits selects from a finite set. Finite sets are even smaller than countable sets. If a random (unknown) quantity has an uncountable number of possible outcomes, for example, the variable can take on any real value between zero and one, then Shannon's results actually show that the outcome *cannot* be represented with a finite number of bits. This is now obvious. Because there are only countably many finite bit sequences, it cannot be possible to use bit sequences to distinguish values from an uncountable set. There just aren't enough bit sequences.

The total number of possible computer programs is $\aleph_0$. The total number of decision functions is bigger, but how much bigger? If it's only slightly bigger, then maybe we haven't lost much by limiting ourselves to what digital computers can do. But it isn't only slightly bigger. It is actually *vastly* bigger.

First, notice that if we add one element to a countably infinite set, the set does not get any larger. For example, suppose we add the element -1 to $\mathbb{N}$ to get the set $\{-1, 0, 1, 2, 3, \cdots\}$. The resulting set has the same size as $\mathbb{N}$, as we can see by this correspondence:

$$
\begin{array}{cccc}
0 & 1 & 2 & 3 \quad \cdots \\
\updownarrow & \updownarrow & \updownarrow & \updownarrow \quad \cdots \\
-1 & 0 & 1 & 2 \quad \cdots
\end{array}
$$

The correspondence includes all elements of both sets.

By the same reasoning, if we add any finite number of elements to a countably infinite set, the size of the set does not change. What if we add a countably infinite number of elements? Suppose, for example, that we add to $\mathbb{N}$ all the elements of $\mathbb{M} = \{-1, -2, -3, \cdots\}$. The combined set is the set of all integers, often written $\mathbb{Z}$. The combined set $\mathbb{Z}$ again has the same size as $\mathbb{N}$! This correspondence again includes all elements of both sets:

$$
\begin{array}{ccccccc}
\mathbb{N}: & 0 & 1 & 2 & 3 & 4 & 5 \quad \cdots \\
 & \updownarrow & \updownarrow & \updownarrow & \updownarrow & \updownarrow & \updownarrow \quad \cdots \\
\mathbb{Z}: & 0 & -1 & 1 & -2 & 2 & -3 \quad \cdots
\end{array}
$$

Intuitively, it would seem that we have doubled the size of the set, but actually we haven't changed the size at all. Thus, an uncountable set is more than twice as large as $\aleph_0$. But it's even much bigger than that.

The set of rational numbers, it turns out, is also countably infinite. A rational number r is any number that can be written as the ratio of two integers n and d (i.e., as n/d). To make it easier to find the correspondence, let's restrict ourselves to only positive n and d. We can then form a table that includes every possible rational number as follows:

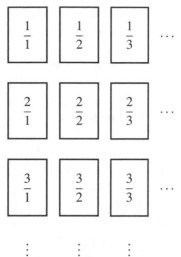

The ellipsis $\cdots$ means simply to continue the pattern horizontally and vertically forever. This table has more entries than there are rational numbers because there are some redundancies. For example, the diagonal elements, 1/1, 2/2, 3/3, and so on, all represent the same rational number 1. But every positive rational number is somewhere in the table.

We can establish a one-to-one correspondence between the entries in this table and $\mathbb{N}$ by traversing the table as shown by the arrows below:

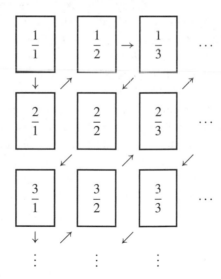

If you now follow the arrows in the table, you can see how to "count" the rational numbers, hitting every single rational number in a well-defined order. Following the path of the arrows and eliminating any redundancies as you go, you can establish a correspondence between the set of all positive rational numbers and the set $\mathbb{N}$. Just start at the upper left and associate 1/1 with the natural number 0. Follow the first arrow and associate 2/1 with the natural number 1. Follow the second arrow and associate 1/2 with the natural number 2. Continuing like this, we can associate every positive rational number with a unique natural number. We can then show that the set of all rational numbers, positive and negative, is also countable, using the same trick we used to show that $\mathbb{Z}$ is countable.

A square array like the table above, if it is finite, is equal in size to the square of the number of rows and columns. Ignoring the ellipsis $\cdots$ in the table above, there are three rows and three columns, for a total of nine entries, the square of three. Letting the table grow, which is what is implied by the ellipsis, the size of the table will become n^2, where n is the number of rows and columns. As the table grows to infinity, the number of rows and columns will become $\aleph_0$, suggesting that the size of the table should be $\aleph_0^2$. However, because the entries in the table are countable, $\aleph_0^2 = \aleph_0$. Thus, intuitively, the size of $\mathbb{R}$ and the size of the set of decision functions is larger than the *square* of $\aleph_0$. Put another way, an uncountable set is not only bigger than two infinite sets with size $\aleph_0$ combined, but it is bigger than the combination of a countably infinite number of countably infinite sets! In fact, it's bigger than any finite power of $\aleph_0$. Hence, the set of decision functions really is *vastly* bigger than $\aleph_0$, and there are even bigger sets that are vastly bigger than the set of decision functions.

An obvious question arises: is there any set whose size lies between $\aleph_0$ and the set of decision functions? Mathematicians usually assume that no such set exists. This hypothesis is called the "continuum hypothesis." It is unproven and in fact cannot be proven. It must be assumed. In 1939, the Austrian-American mathematician Kurt Gödel proved that the continuum hypothesis cannot be *disproved* using the accepted axioms of set theory.[7] In 1963, the American mathematician Paul Cohen proved that the continuum hypothesis also cannot be proved from these same axioms. The continuum hypothesis is therefore independent of the axioms of set theory. But regardless of whether we assume the continuum hypothesis, it remains true that any uncountable set has vastly more elements than any countable set.

The proof I gave above that computers cannot solve all decision functions offered just one counterexample, a single decision function C that could not be implemented by any program. The same argument proves that no countable set of programs can realize all decision functions. So this shows that the size of the set of decision

---

7 Specifically, using the axioms of Zermelo-Fraenkel set theory, from which much of mathematics can be derived.

functions is *strictly* larger than the set of programs. The set of decision functions is uncountable and therefore *vastly* larger than the set of decision functions that can be computed by any computer.

Turing's result, that decision functions exist that are not effectively computable, is one of several results emerging around the same time that crushed the optimism of the previous century.[8] In the face of such results, particularly when viewed through the lens of cardinality, I have to conclude that it is extremely improbable that every interesting information-processing machine is somehow a piece of software. To believe something so improbable in the face of such weak evidence requires a great deal of faith. In chapter 11, I will show how to systematically use evidence to update our beliefs (using Bayesian reasoning) and why an improbable hypothesis demands stronger evidence.

Fortunately, engineers are not limited to working with software. As described in chapter 6, cyberphysical systems form a partnership between software and other nonsoftware physical systems. These combined machines offer a vast and largely unexplored landscape for creative designers and inventors. Even more interesting, the partnership between computers and humans has vastly more potential than either alone, as I will argue in the next chapter.

8.4   Digital Physics?

Digital physics postulates that nature does not and cannot have a continuous range of possibilities, the total number of possible states that any system can have (including the entire universe) is finite, and physical systems are essentially equivalent to software. Digital physics is a paradigm shift, in the sense of Kuhn, and I hope I am not just one of those opponents who must eventually die so the paradigm can become universally accepted.

If it is true, then the postulate has severe consequences. It means that many of our most cherished ideas of the physical world are wrong, including that space is a three-dimensional continuum and time progresses fluidly from one instant to the next. It means that Newton's laws and Einstein's relativity are are both wrong because both depend on time and space continuums. Of course, as stated by Box and Draper (1987), all models are wrong, but some are useful. By this principle, digital physics must also be wrong. It is a model, a map, not a territory.

For most purposes, digital physics is unlikely to be as useful as models that admit continuity in the physical world. Even if it *is* finite, the number of states of all but the most tiny systems will be enormous compared with what digital technology can manage today and in the foreseeable future. Nevertheless, the idea of digital

---

8 For a wonderful account of this optimism and its downfall, see Kline (1980).

physics provokes deep questions about modeling from scientific, engineering, and philosophical perspectives.

From the engineering perspective, digital physics postulates that everything that is possible to make, in principle, can be made with software. It means that dishwashers are, in fact, information-processing machines. It means that the human mind and all of its cognitive functions are, in principle, realizable in software (I will return to this question in section 9.3 in the next chapter). It means that no machine can accomplish what software cannot do, such as computing undecidable functions or working with real numbers. Hence, it is pointless to try to build such machines.

From the scientific perspective, digital physics postulates that nature is extremely constrained, operating within a tiny subset of mechanisms that might have been possible. This tiny subset includes almost none of the physical theories that humans have developed over centuries.

From a philosophical perspective, if the number of states of the universe is finite, then it must be true that (1) the universe must eventually find itself in a state that it has been in before; (2) the universe can only change state a finite number of times in infinite time, so it must effectively stop changing; or (3) time must end. I find all of these possibilities profoundly disturbing.

Perhaps even more disturbing is that it may be impossible to disprove digital physics. Specifically, if we assume that all measurements of the physical world are noisy, then the Shannon channel capacity theorem, equation (4), states that every measurement will convey only a finite number of bits of information. Therefore, any measurement that attempts to show that the number of states of some system is infinite will fail. Hence, by Popper's philosophy of science, digital physics is not scientific because it is not falsifiable. Digital physics becomes a faith.

Digital physics strikes me as far fetched, but most of modern physics is. Most of physics in the universe operates in regimes where our human senses and the intuitions built through them are useless. Our senses and our experience on earth do not give us much intuition to use in understanding black holes and quarks. So by now we should be used to counterintuitive theories. But there is one aspect of digital physics that makes it highly improbable and, hence, extremely surprising. It constrains the universe to operating within a countable and, worse, a finite system. Given that almost all of our deepest and best understanding of the world has come from powerful models of continuums and infinities, for example, Newton's and Leibniz's calculus, to conclude that nature has no need for anything infinite gives pause. It seems like a throwback to preenlightenment days. At a minimum, we should insist on incontrovertible evidence before accepting this. Although the evidence today looks weak to me, quite a consensus has formed among physicists supporting this hypothesis, so the evidence must not look weak to them. In chapter 11, I will explain exactly what I mean by "evidence," but for now let's just examine what digital physics means and why it is so improbable.

Digital physics has several variants, some of which are bizarre. The variants can be put in order from weakest to strongest as follows:

1. In its weakest form (fewest assumptions), digital physics asserts that the number of possible states of any system in nature with finite energy and volume is finite. If this is the case, then the state of any such system can be completely encoded with a finite number of bits.

2. A slightly stronger form asserts that the physical world is essentially informational. Every process is an information transformation, and every object in the world is essentially a bundle of information. Digital physics further asserts that information can be measured in bits.

3. In a stronger form, digital physics assumes that every physical process is essentially a computation, representable in principle as software. This requires that processes in nature be algorithmic, proceeding as step-by-step operations.

4. In a still stronger form, digital physics assumes that the physical world is essentially a computer.

5. In the strongest form that I have seen, digital physics asserts that the physical world is a simulation carried out by a computer.

In my opinion, these philosophies are confusing the map with the territory. Are they talking about models of reality or about reality?[9]

A supporter of at least the weaker forms of digital physics was the Mexico-born Israeli-American theoretical physicist Jacob Bekenstein, who was a professor at the Ben-Gurion University and then the Hebrew University in Israel until his unexpected death in 2015. Bekenstein and his colleagues developed what is now called the "Bekenstein bound," an upper limit on the entropy that can be contained within a given finite volume of space that has a finite amount of energy (see Freiberger [2014] for a short readable summary). If we assume that the form of entropy that Bekenstein considered is discrete entropy, explained in the previous chapter, then the Bekenstein bound shows that the amount of information, measured in bits, that can be stored in a given volume of space is limited. Equivalently, the bound shows that anything occupying a given volume of space can be completely described, down to the quantum level, with a finite number of bits. I will call this the "digital interpretation of Bekenstein's bound." Under this interpretation, the first form of digital physics listed previously follows immediately. An alternative nondigital interpretation of the Bekenstein bound using continuous entropy appears to be consistent with Bekenstein's original formulation (Bekenstein, 1973), but this is not the interpretation adopted by most physicists today.

---

9 The existence of a reality independent of humans is not a universally accepted truth. Philosophers call this assumption "realism," and for the purposes of this argument, it is a position I will adopt.

So under the digital interpretation, how many bits can we store in a given space? James Redford, who also claims that modern physics proves the existence of God, in a 2012 paper uses the Bekenstein bound to calculate the number of bits required to encode a human being (Redford, 2012, p. 126). He concludes that an adult human male can be encoded in 2×10^{45} bits. My laptop's hard disk can store one terabyte, or 10^{12} bytes, or roughly 10^{13} bits. So I would need 10^{32} such laptops to store this many bits. This is 10^{32}:

$$100,000,000,000,000,000,000,000,000,000,000.$$

If Moore's law continues unabated and we assume it applies to memory storage devices, then in only about 130 years, we may have a computer with this much memory. Such is the power of Moore's law, which predicts a doubling of the number of transistors on a chip every two years.

Although the digital interpretation of Bekenstein's bound seems to be widely accepted among physicists today, I remain stoically skeptical, an admittedly lonely position. I do not doubt Bekenstein's result that entropy in a volume of space is limited and finite. What I doubt is that Bekenstein's entropy is discrete entropy, and hence represents information that can be encoded in bits. Bekenstein's arguments seem to work just as well using continuous entropy, explained in section 7.4 of the previous chapter. It is a mistake to give a digital interpretation to continuous entropy. The continuous entropy in a system does not tell us how many bits it takes to encode that system, although it *does* quantify information content. If we are using continuous entropy, the system *cannot* be encoded with a finite number of bits. The system nevertheless has information, and its information content can be compared with the information content in other continuous systems.

Ludwig Boltzmann and his contemporaries defined the thermodynamic entropy in a macroscopic system, such as a volume of gas, by the formula $k \log(M)$, equation (32) on page 133. In the usual explanations, M is the number of states that the microscopic system, consisting of the individual molecules in the gas, can be in such that the macroscopic state (volume, pressure, mass, and temperature) are as observed. The scaling constant k is called the Boltzmann constant, and it simply changes the units that we are using to measure entropy. This explanation assumes that the M states are all equally likely and that *the number of such states is finite*. To interpret this as equivalent to bits in discrete entropy, we have to assume that molecules can only have a finite number M of possible states. This assumption is a form of digital physics. Hence, to interpret thermodynamic entropy as a measure of information in bits, we have to first assume that digital physics is true. To then use the Bekenstein bound to prove digital physics is a logic error. There may be *other* reasons that the number of states is finite, but the reason cannot be that the entropy is finite.

The error here is subtle but important. Boltzmann couldn't have known how many possible states the microsystem could have that were consistent with an observed macrostate. Boltzmann wanted to model the state of a molecule by its position and velocity (or momentum). In Boltzmann's time, these would have been continuous random quantities, so the entropy should be more properly interpreted as continuous entropy. Today, the number of possible states is understood to be determined by quantum mechanics, which had not been developed in Boltzmann's day. In an alternative explanation, M is a stand-in for *relative* degrees of freedom. It was common in classical thermodynamics, before quantum mechanics, to replace M with a number proportional to the number of molecules in the volume of gas being considered. In this case, the entropy has an arbitrary offset, but as long as the same offset is used for any two entropies, it remains valid to *compare* entropies. But the absolute value of the entropy loses any physical meaning.

After Boltzmann, formulas for entropy have been improved by physicists to take into account more knowledge of the underlying physics, including quantum effects, and thereby acquire more direct physical meaning. An example is the so-called Sackur-Tetrode equation for the entropy in an ideal gas, derived in the early 1900s, which has both classical and quantum mechanical aspects. I will spare you the details, but Wikipedia provides a starting point if you are interested. This equation, however, must be a continuous entropy measure, not discrete entropy, because it does not constrain the entropy to be positive, and it sets the entropy at minus infinity when the temperature gets to absolute zero. A physicist would likely tell you that the equation becomes invalid at low temperatures, where the approximations used to derive the equation are no longer accurate. This may be true, but if this is a continuous entropy, then it is a mistake to read it as a number of bits of information at *any* temperature. This landscape underscores the difficulty in keeping straight whether a discussion of entropy is talking about discrete or continuous entropy. The two are not comparable.

If in fact a physical system has an infinite number of possible states consistent with the observations, and if the probability density function for these states is known, then in fact we can define a continuous entropy for the system, and the number *does* have physical meaning, as shown in section 7.4 of the previous chapter. This entropy quantifies information content, just like discrete entropy, but the information *cannot* be encoded in bits. An infinite number of bits would be required, just as an infinite number of bits is required to encode a real number, even though the number is finite. If, further, the probability density function is uniform, as depicted in figure 7.1, then the form of expression for continuous entropy will be exactly $k \log_2(M)$, where M now is simply the height of the probability density function. So Boltzmann's formulation works fine even if the number of states is *not* finite, as long as all the states are equally likely.

Entropy is the central concept in the second law of thermodynamics, which states that entropy increases in any system (or at least does not decrease). So the second law of thermodynamics is all about *comparing* entropies, and not at all about their absolute values, so it is completely unaffected by whether we interpret entropy as a measure of information in bits. It is even unaffected by the arbitrary offset that results if we do not know the probability density function for the states. The second law of thermodynamics works just as well with continuous entropy as with discrete entropy. To give a digital interpretation to entropy, to measure it in units of bits, we need to assume that the number of possible states of a molecule is *finite*. We have to assume digital physics.

Too many physicists seem to assume that the word "entropy" automatically means discrete entropy. Seth Lloyd, a professor of mechanical engineering and physics at MIT, in his 2006 book *Programming the Universe*, does this repeatedly. He says about the second law,

> It states that each physical system contains a certain number of bits of information—both invisible information (or entropy) and visible information—and that the physical dynamics that process and transform that information never decrease that total number of bits. (Lloyd, 2006)

But the second law works absolutely unmodified if the underlying random processes are continuous, in which case the information is not representable in bits. Other principles in physics may result in bits, and Lloyd argues strongly that quantum mechanics does this. However, a physical system can have a finite entropy and still not be representable in bits if that finite entropy is a continuous entropy.

Lloyd defines entropy digitally, saying, "entropy is a measure of the number of bits of unavailable information registered by the atoms and molecules that make up the world." And "the quantity called entropy is proportional to the number of bits required to describe the way atoms are jiggling." But continuous entropy is still entropy, and it is not a measure in bits. Lloyd then makes an extraordinary claim that "as the statistical mechanicians of the late nineteenth century showed, the world is made up of bits." Those statistical mechanicians, Boltzmann and his contemporaries, had never heard of bits, and their theories work fine with continuous entropy.

Lloyd has *other* reasons for assuming that the physical world is digital. In *Programming the Universe*, he argues that the universe is in fact a type of computer known as a "quantum computer." Although the relationship between quantum computers and Turing computation is far from trivial, Lloyd's position is strongly in favor of digital physics at least to the level that everything in the world is digital. However, it is highly misleading to claim that a finite entropy implies that the world is digital.

Quantum computers are based on quantum mechanics, which emerged well after the work of the statistical mechanicians of the late nineteenth century. Quantum mechanics replaces the notions of position and momentum, which those mechanicians would have used, with a "wave function," which relates position and momentum probabilistically. The wave function involves continuous variables and has an offset in space, effectively a position. Is the number of possible offsets of the wave function finite?

Digital physics rests on a principle that for any physical system with well-defined boundary conditions, only a finite number of wave functions that constitute the system are possible. But how to model these boundary conditions is quite subtle. If we are talking about a chamber of gas, as Boltzmann was, then aren't the walls of the chamber properly modeled as wave functions interacting with those of the gas molecules? If so, then the enclosure must be considered part of the system rather than a boundary condition, but so must the environment around the enclosure, and the environment around that. If we assume that ultimately the universe is finite in extent and age, then we eventually get well-defined boundary conditions. But it boggles my mind to rely on the finite scale of the universe to describe behaviors at the subatomic scale where quantum mechanics applies. Even the tiniest approximations or the use of statistical arguments in any calculations that span such a range of values would invalidate the conclusions, and entropy calculations are rife with approximations. But the physicists know more about this than I do, so I will have to take their word for it. Ultimately, the conclusion that the number of states is actually finite seems to depend on models that I do not understand, have not been validated experimentally, and cannot be falsified.

Shannon did show that if an *observation* of a continuous variable is imperfect (it is noisy), then the information *conveyed* by the observation can be measured in bits and is finite. This is the channel capacity theorem, equation (4) in the previous chapter. However, the quantity of information observed is *not* equal to the entropy in that continuous variable, which is finite but not measured in bits. The entropy of a continuous variable is a continuous entropy. The fact that a noisy observation of a continuous random variable conveys only a finite number of bits is a consequence of the *relative* entropy before and after observation. This was Shannon's most profound observation. He showed that the information capacity of any noisy communication channel is finite and measurable in bits.

Nevertheless, perhaps digital physics can be salvaged by assuming that observations are always noisy and the information *conveyed* is all the information that is relevant. Shannon did not show that the information *contained* in the variable is finite, only that the information *conveyed* is finite. In fact, the information contained would require an infinite number of bits to encode. Shannon showed that it is *not possible* to perfectly reconstruct the value of a continuous variable from a noisy observation. To salvage digital physics, we could assume that any information that fails to be conveyed by a noisy observation is not relevant information. This is equivalent to assuming digital physics. We have to assume digital physics to prove digital physics.

This question of information *conveyed* versus information *contained* is a deep and difficult question. Is it possible for a physical system to have information that is not externally observable at all but is nonetheless essential to the system? I briefly addressed this question in section 8.2, where I lamented that, as a teacher, I am unable to convey all the information I carry in my brain. I will return to this question when I consider the human brain and cognition in chapter 9, but ultimately, I believe that even information that is not conveyed is still relevant.

The way Bekenstein developed his bound is an interesting story. He worked with physicist Stephen Hawking on a problem that black holes seem to defy the second law of thermodynamics by swallowing up entropy. To salvage the second law of thermodynamics, Bekenstein and Hawking associated the surface area of the event horizon of a black hole with entropy. Nobody before Bekenstein and Hawking had thought that a surface area of anything had anything to do with entropy, but this association resolved the problem, saving the second law.

The bound also depends on the theory of quantum gravity, an effort to reconcile quantum mechanics with Einstein's general theory of relativity. Neither this nor equating surface area with entropy is without controversy, and neither has had any experimental observation. For me, accepting Bekenstein's bound requires a great deal of faith in physics that is difficult if not impossible to fully understand and may be beyond the reach of experimental observation. Even Hawking, one of the most widely recognized and respected physicists today, expresses doubt, pointing out that the digital interpretation of the bound is inconsistent with most of modern physics. Hawking notes that continuums are required in both time and space for the bedrock of quantum mechanics, the Schrödinger equation, resulting in "an infinite density of information which is not allowed" (Hawking, 2002). Unless you *assume* that the number of possible states is finite, the Bekenstein bound talks about continuous entropy, not entropy in bits, and the contradiction evaporates.

At this time, there is no experimental confirmation of a digital and quantized nature of the universe, both needed for digital physics. Experiments in progress are looking for such confirmation. One is the Fermilab Holometer near Chicago that is intended to be the world's most sensitive laser interferometer, more sensitive than LIGO (see chapter 1). According to Wikipedia, the principal investigator on this project, Craig Hogan, states,

> We're trying to detect the smallest unit in the universe. This is really great fun, a sort of old-fashioned physics experiment where you don't know what the result will be. [Wikipedia page on Holometer, Retrieved May 24, 2016]

The experiment started collecting data in August 2014, and as of August 2016 has no major results yet. Worse, even if it gets results, if there is any noise at all in the measurements, then Shannon's channel capacity theorem tells us that the measurements can only convey a finite number of bits of information, even if the underlying system has an infinite number of bits of information.

I am not a physicist, so you should take my skepticism with a grain of salt, but I have to say I would be extraordinarily surprised if digital physics is valid. It reads more like a cult than a science to me. If it is valid, then nature has its hands tied indeed. For some reason, nature has restricted itself to operating within only a finite number of possibilities. Why would nature do that? Although my skepticism seems to be a minority opinion, I am not *entirely* alone.

Sir Roger Penrose, an English physicist, mathematician, and philosopher, in his controversial book *The Emperor's New Mind*, states,

> The belief seems to be widespread that, indeed, "everything is a digital computer." It is my intention, in this book, to try to show why, and perhaps how, this need not be the case. (Penrose, 1989, p. 30)

Penrose goes on to argue that consciousness, a naturally occurring process in the physical world, is not only not a computation but is not even explainable using the known laws of physics.

Gualtiero Piccinini, a philosopher at the University of Missouri, observes,

> $\cdots$ from the point of view of strict mathematical description, the thesis that everything is a computing system $\cdots$ cannot be supported. (Piccinini, 2007)

Piccinini, like me, defends this idea using the notion of cardinality. There just aren't enough possible computations to encompass the richness of the physical world.

Digital physics cannot be disproved, assuming all measurements have noise, so this issue may never be resolved. It will probably always remain a matter of faith. My faith is that nature is more likely to be richer in possibilities than poorer. The tiny cardinality of a digital universe just seems too small to me.

9 Symbiosis

··· in which I go beyond the countable world of computing and argue that computers are not universal machines and their real power comes from their partnership with humans. I explain the notion of a continuum, a concept that is out of reach for software and rejected by digital physics but seemingly essential for modeling the physical world; I argue that computers are co-evolving symbiotically with humans; and I examine the limitations of the formal models that underlie what computers do, showing that the partnership between humans and computers is much more powerful than either alone.

9.1 The Notion of a Continuum

In chapter 8, I mentioned the continuum hypothesis, an unproven (and in some sense unprovable) assertion that the next larger infinite sets larger than the natural numbers $\mathbb{N}$ have the same cardinality (size) as the set $\mathbb{R}$ of real numbers. I also mentioned that $\mathbb{R}$ has the same size as the set of all decision functions.

The word "continuum" in "continuum hypothesis" is an interesting and deeply philosophical one. What is a continuum? Intuitively, a continuum is a set where one can move "smoothly" from one element in the set to another in the set without having to pass through any values not in the set.

The set of real numbers is a continuum. Suppose I want to move from real number 3 to real number 4. Every number between 3 and 4 is a real number, so as I pass through all those numbers in between, I never leave the set of real numbers.

The set of rational numbers is not a continuum. Consider again moving from 3 to 4. The numbers 3 and 4 are both rational numbers, so they are in the set of rational numbers. But if I try to move smoothly from 3 to 4, I would have to pass through $\pi \approx 3.14159\cdots$. However, π is not a rational number, so to move smoothly from 3 to 4, I have to leave the set of rational numbers.[1] The set of rational numbers is not

---

1 One way to construct the set of real numbers from the set of rational numbers is using the idea of a Dedekind cut, named after the German mathematician Richard Dedekind. A Dedekind cut is a partition of the set of rational numbers into two nonoverlapping subsets A and B, where all elements of A are less than all elements of B and A has no greatest element. The set of all such cuts can be put into a one-to-one correspondence with the real numbers and in fact can be taken to *define* the set of real numbers. The set B may or may not have a least element. If B has a least element, then the cut is put into correspondence with that least element, a rational number, and hence also a real number. If B has no least element, then the cut is put into correspondence with the unique irrational number that lies between A and B, in a sense filling the "gap" between them. This gap prevents the set of rational numbers from forming a continuum. It is remarkable that the number of such cuts, the cardinality of the set of Dedekind cuts, is vastly larger than the number of rational numbers, a result due to Cantor.

a continuum. In fact, no countable set is a continuum. Because computers operate entirely within a realm of countable sets, continuums are out of reach for computers.

Now we can get philosophical. Do continuums exist in the physical world? Consider time as a set. Let's just focus on the possible times in one afternoon. Suppose it is now 3 PM, a member of that set. As I move from 3 PM to 4 PM, do I pass through any instants that are not times? I have to assume not. If I further assume that time progresses "smoothly," then I would have to conclude that time is a continuum. Although most models in physics assume that time is a continuum, some modern physics contests this assumption.

The physicist John Archibald Wheeler, who coined the term "black hole" and was Bekenstein's PhD thesis advisor at Princeton, wrote the following:

> Time, among all concepts in the world of physics, puts up the greatest resistance to being dethroned from [the] ideal continuum to the world of the discrete, of information, of bits. (Wheeler, 1986)

Wheeler can't see any way to handle time without continuums. Despite these reservations, Wheeler was a proponent of digital physics, which requires that continuums not exist in the physical world. Wheeler even coined the phrase "it from bit" to capture the essence of digital physics.

Space seems to similarly require a notion of a continuum. If I am standing at point x and I move to point y, do I pass through any points that are not in space? Do I move smoothly? Although my senses seem to indicate that I do, I cannot trust what my senses tell me when I am reasoning about time and space scales smaller than what my senses can perceive. Nevertheless, almost all of physics models space as a continuum.[2]

Of course, we can now go down the rabbit hole to debate what "smoothly" means, but I will instead rest on centuries of tradition in science, where time and space are nearly universally modeled as continuums. At a minimum, we have to concede that modeling time and space as continuums has proved to be a useful paradigm indeed. These are just models, of course, so as we faithfully avoid confusing the map and territory, we cannot assert the existence of continuums in the physical world just because they are useful as models. However, the reason that they are useful as models is that these models provide *simpler explanations* of the physical world than models that reject continuums. Applying the principle of Occam's razor, attributed to William of Ockham (c. 1287–1347), an English Franciscan friar, scholastic

---

2 An exception is the holographic principle, a property of string theories and possibly of quantum gravity first proposed by the Dutch physicist Gerard 't Hooft. The holographic principle replaces the notion of three-dimensional space with lower dimensional surfaces and in at least one form replaces quantum theory with a new deterministic theory. There is quite a bit of controversy around this and related theories (Smolin, 2006).

philosopher, and theologian, when there are competing hypotheses, other things being equal, we should choose the simpler one.

We might just as well question whether the physical world has any intrinsic notion of integers or, more broadly, countable sets. Suppose I have two apples sitting in front of me. It certainly seems that these apples are distinct, individual, integral objects. From that observation, I can argue that integer arithmetic, like $1 + 1 = 2$, is a physical reality. But is it? Some of the apple molecules have actually escaped into the air, so I can smell them. What if I take a knife and peel a bit off one apple? Do I still have two apples? What if a worm has eaten part of one? What if the two apples are touching one another? Where does one end and the other begin? Arguably, the notion that these apples are integers is confusing the map and the territory. To *model* them using integers is defensible on the grounds of its usefulness. The grocery store may count the apples in my basket to determine how much to charge me. But aren't integers here just a model? The grocery store could equally well charge me by their weight.

Leopold Kronecker, a nineteenth-century German mathematician, vehemently criticized Georg Cantor's work, famously stating, "God made the integers; the rest is the work of man." If I substitute "nature" for "God," then I would have to conclude that Kronecker could equally well have gotten it exactly backward! It is just as defensible to say, "Man made the integers; the rest is the work of God." Software, one of the most remarkable human constructions ever, is all about integers and only about integers, and even then only about a tiny subset of all conceivable operations on integers. It is not unreasonable to expect the natural world to be vastly richer than that.

If we further assume that the statement "time is a continuum" is a Platonic truth about the physical world, then we have to assume that physical systems that deal with continuums exist. I am uncomfortable doing this because once again it risks confusing the map and the territory. The notion of a continuum is a mathematical model not a physical reality. But software systems also operate in the world of models not in the physical world. We know how to make physical machines (computers) that are faithful to the software model. Can we make physical machines that are faithful to models in a continuum? Yes, we can! Mechanical engineers, for example, do it all the time. My dishwasher is a machine that almost certainly is better modeled using continuums than restricting it to computation.

Computers have two key limitations. First, they operate only on digital data, limiting their domain to a countable set. Second, their operations are algorithmic, performed as a sequence of steps, where time is irrelevant. Unless you subscribe to a strong form of digital physics, the physical world has neither of these constraints. The models of resistors and inductors considered in chapter 2, for example, have neither of these properties, yet these are arguably models of machines.

But all models are wrong, so it could be that the models we use for resistors and inductors are not only wrong in the ways I cited in chapter 2 but also wrong to be operating in a continuum and wrong because they do not describe the behavior algorithmically, as a sequence of steps. The digital physics advocates would have to conclude that these models are wrong in this way. To me, however, an algorithmic model of a resistor or an inductor operating on integers only would be cumbersome, useful only for computer simulation and opaque to people.

The fact that computers do not and cannot deal with continuums in no way proves that there are no machines that *can* deal with continuums. Moreover, continuums are not the only larger infinite sets bigger than the countable sets. Some sets are vastly larger than continuums, and other sets are vastly larger than those sets. Why would nature limit itself to the smallest of infinite sets? This notion seems so improbable that it requires compelling evidence before accepting it. I've already argued that because of the Shannon channel capacity theorem, we cannot obtain such evidence by empirical measurement. In the next section, I will argue that physical systems, including information-processing systems such as the human brain, are far more *likely* to be machines that deal with uncountable sets than they are to be computers.

9.2 The Impossible Becomes Possible

Consider a simple balloon. I can think of this balloon as a machine that outputs the circumference of a circle given its diameter. Suppose I inflate the ballon until it reaches a specific diameter d at its widest point. The balloon then exhibits a circumference at a cross-section at that widest point. The "input" to the machine is the diameter d, and the output is the circumference, which by the basic geometry of circles should come out to $\pi \times d$. If I inflate the machine to a diameter of one foot, it calculates π.

The number π is not representable as a binary number with a finite number of bits. So a digital computer would have to run forever before it could produce as output a binary representation of π. As it happens, the number π is nevertheless "computable," in that, given any positive integer n, a computer can calculate the nth digit of π. No computer can give us *all* the digits of π in finite time, but a computer can give us any arbitrary digit of a decimal or binary representation of π.

Equivalently, there is a *finite* program that, in effect, describes the infinite sequence of digits that constitute π. Because this finite program "describes" π, this infinite sequence is "describable" (by a finite description). However, there are many more real numbers where there is no computer program that can give us any arbitrary digit of the number. Gregory Chaitin, an Argentine-American mathematician who worked at IBM in New York and at the University of Auckland, New Zealand, developed a beautiful example of such a number, one that he called "Omega," or Ω. Ω is a number between zero and one whose binary representation can be used to solve Turing's halting problem for a particular binary encoding of Turing machines. Specifically, if we know the first N bits of the binary representation of Ω, then we can determine for all valid programs of length up to N bits whether they halt. Because this question is known to be undecidable, no computer program can give us any arbitrary bit of the binary representation of Ω.[3]

My balloon machine is not trying to calculate Ω, but it nevertheless does seem to do something no computer can do. Specifically, it outputs, all at once, a representation of π.

I'm sure that you are protesting now. Any reader who has persisted this far in this book is far too smart to be hoodwinked by this slight of hand. This argument has several problems. First, the circumference of the balloon will not be π because the balloon is not a Platonic Ideal. It has imperfections. The rubber is probably not perfectly uniform, so the balloon will not form a perfect circle. Thus, the circumference will be something other than $\pi \times d$. In fact, if we are lucky, it could be $4 \times \Omega \times d$, in which case we have built a machine that calculates Ω.

Nevertheless, I stick to my guns. Yes, the balloon will not form a perfect circle, but it will form some shape. If we assume that this shape has a circumference, then it is likely that circumference will be a noncomputable multiple of the diameter. There are vastly more noncomputable numbers than computable numbers, so it would require digital physics to assume that the actual circumference of the balloon is a computable multiple of the diameter. I don't actually know what function the balloon realizes, but it seems that it most likely realizes a function that no digital computer can realize.

But wait. This argument has still more problems. Suppose that the input to the balloon machine is restricted to a countable set perhaps because we accept digital physics. Then the set of all possible outputs is also countable. We know from the arguments in the previous chapter that a union of two countable sets is still countable. Thus, if the inputs are countable, then the balloon machine is actually working only with countable sets.

Suppose that the balloon is perfect, in that given any input diameter d, its circumference will be exactly $\pi \times d$. Then I can easily come up with a binary encoding that handles all the inputs and outputs of this machine if d comes from a countable set. For example, suppose that my binary encoding is such that any sequence of bits that begins with zero is interpreted as an integer, where the bits after the first zero directly encode that integer. For example, 00 means zero, 01 means one, 010 means two, 011 means three, and so on. Suppose further that in my binary encoding, any sequence of bits that begins with 1 is interpreted to mean $\pi \times n$, where n is the binary number following the leading 1. For example, 10 means $\pi \times 0 = 0$, 11 means π, 110 means 2π, and so on. I now have an encoding that a computer can use to realize the balloon machine.

---

3   For a wonderfully readable story about Ω, see Chaitin (2005).

Computer scientists make a distinction between *syntax*, the way things are written, and *semantics*, what things mean. A computer transforms bit patterns, operating only on syntax, and is restricted to a countable set of syntactic objects, bit sequences. However, a human looking at those bit sequences is not restricted to a countable number of *interpretations* of those objects. To interpret 110 to mean 2π is a human act not something the computer does. A human assigns semantics to the syntax 110. Bit sequences can mean integers, real numbers, text, anything, in fact. We can even encode emotions. I can declare that the bit sequence 01010 means "happy" and write a program that produces "happy" as its output. Does this mean that the computer is happy?

Semantics is an association between a set of syntactic objects, such as bit sequences, and a set of concepts. Numbers are concepts, so one possible semantic interpretation of a sequence of bits is as a binary number. However, many other interpretations are possible. In fact, there is no reason to assume that the number of possible interpretations is countable. What makes computers so effective, so useful to humans, is the many possible interpretations we can assign to bit sequences. The partnership of computers with humans is the real source of their power.

But what if humans are just computers? If we assume a strong form of digital physics, then this must be true, ultimately. Even if we do not assume digital physics, many smart people believe that the human mind is in fact software. Alan Turing was a strong advocate of this point of view. In this case, semantics must somehow be reducible to syntax, and the set of all semantic interpretations must be countable.

This idea brings us back to the question of whether it is possible to realize machines that do more than what Turing defined as "computation." If it is not possible, then the human brain, a machine, must be performing computation. It must, therefore, be limited to operating within a countable world. But if it *is* possible, then the world is much richer, and there are no bounds to creativity.

Returning to our balloon machine, suppose that I inflate the balloon to a diameter of $d = 1$ foot. What is the output circumference from the machine? In fact, what is the input? How do we know it is exactly one foot? If we want numbers, then we have a problem that is potentially of the scale of the LIGO problem. Perhaps we write a proposal for $1.1 billion to the National Science Foundation to fund a research project to make this measurement. Presumably, if we get funded, we could enlist the 1,019 scientists and engineers who worked on the LIGO project to measure the diameter and circumference of the balloon to precisions much less than the diameter of a proton. But this is just representing the information in another form. The information is already represented in a perfectly adequate and much cheaper form in the balloon. But the problem with that form is that we humans don't know what the input is, what the output is, nor what function the machine computes!

It is easy to assign semantics to bit sequences. Suppose I declare that 010101 means "happy" and 111000 means "sunshine." Suppose further I have a computer running a program that, given 111000 as input, produces 010101 as output. This computer outputs "happy" in response to the input "sunshine." This is simply true by definition. However, the balloon machine is more problematic. I don't know the input, the output, or the function being computed.

What good is a machine if we can't know the function that it realizes? My claim is that we should not be surprised that we cannot know the function that it realizes. We might assume that to "know" the function means that we can describe that function in some mathematical or natural language.[4] Given any written mathematical or natural language, the vast majority of functions with numerical input and output are not describable in that language. This is because every description in any such language is a sequence of characters from a finite alphabet of characters, so the total number of such descriptions is countable. There are vastly more functions, so there can't possibly be descriptions for all of them in any one language. In fact, any language will only be able to describe a tiny subset of them, a countably infinite subset. Does a function need to be describable to be useful?

A car is a machine whose function is to carry me some distance. I don't need to measure that distance to make use of the machine. In fact, any measurement of the circumference of the balloon is simply putting the output information into another form. What's wrong with the original form given to me by the balloon? If I insist on writing the circumference down as a decimal number on paper, then I have already forced the problem into the realm of digital computers.

Moreover, if I assume that every measurement is noisy, then by Shannon's channel capacity theorem, equation (4) in chapter 7, the measurement will only reveal a finite number of bits of information, even if the underlying physical system contains more information than that. But I don't need to measure or write down the distance that my car carries me to get value out of it. I don't need to measure or write down the circumference of the balloon to assert that the circumference has been produced by the balloon machine.

A rather simple and much more practical example of a machine that is not implementable in software is a simple inductor, described in chapter 2. Assume that the input to the machine is the voltage and the output is the current. The relationship between the input and output is given by Faraday's law, equation (256) on page 43. Under Faraday's law, this inductor implements a function that is not effectively computable and therefore not implementable in software. This is true simply because the input and output both exist in a continuum; even if the input is drawn from a countable set, under Faraday's law, the outputs over time have an uncountable number of values.

---

4 The notion that "knowing" something is not the same as being able to describe it was immortalized in 1964 by United States Supreme Court Justice Potter Stewart who described his threshold test for obscenity in Jacobellis v. Ohio as, "I know it when I see it."

To assert that an inductor is not an information-processing machine, we could try to assert that it is invalid to represent information using a voltage or current. Because all computers represent information using voltages and currents, our whole world of information-processing machines would collapse. Or we could assert that a value in a continuum does not represent information because it cannot be encoded with a finite number of bits. As I explained in chapter 7, it is perfectly valid to interpret a value from a continuum as information.

To assert that an inductor is actually a Turing computation, we would have to reject Faraday's law, start counting electrons, and discretize time. This will not lead to a useful model.

As I pointed out in chapter 2, an inductor is not actually implementable. But just like my balloon, any physical device that I call an inductor is actually a machine that reacts to input voltage by producing a current. Just because I don't have an exact model for that machine doesn't mean the machine doesn't exist. It is an information-processing machine, I just don't know exactly what function it implements.

Even if I did know the input-output function that my machine implements, and even if I can arbitrarily closely approximate that function with a computer, I *still* cannot conclude that I've somehow captured all important properties of the machine. The relationships between the inputs and outputs may not be sufficient. I examine this issue next.

9.3 Digital Psyche?

What good is a machine if we can't know its output or even the function that it computes? I will now give a real-world example of an extremely useful information-processing machine that has properties not observable from outside the machine and has functions that are probably not describable: the human brain. One of the functions that the brain performs is to create consciousness. I know this for a fact because I have a brain, and what we mean by "consciousness" is exactly what I experience as consciousness. In Searle's words, "the concept that names the phenomenon is itself a constituent of the phenomenon."

However, the consciousness that my brain produces is not directly observable to anyone but me. It is a property of my brain, like the circumference of the balloon, and any attempt to externally measure it will fail to capture it. Regardless, I know for a fact that it exists, observable or not. I will not accept any argument that it does not exist. *Cogito ergo sum*, "I think, therefore I am," to quote the seventeenth-century French philosopher, mathematician, and scientist René Descartes. To deny that my consciousness exists would be to deny existence. If we deny that properties not externally observable are important, then we would be forced to conclude that consciousness is not an important property of the human brain. I'm not willing to do that.

Consciousness is one of many cognitive functions of the brain, along with understanding, reasoning, learning, sentience, and remembering. Of these functions, reasoning seems closest to computation. The human brain is clearly capable of some modest form of Turing computation. We can, in our heads, perform the same functions as the logic gates discussed in chapter 4. We have memory, and we are able to follow recipes, step-by-step procedures, emulating a computer executing a program. But we are not actually good at this kind of computation, at least not when we are doing it consciously. A computer performs the logic functions of chapter 4 billions of time per second and stores billions of bytes in memory. We don't even come close.

The human brain does many things that we do not know to be Turing computation. Consider face recognition, something that computers are only now starting to get good at. The fact that computers are starting to excel at face recognition, natural language understanding, and speech recognition leads many engineers and scientists today to conclude that the human brain must be accomplishing these things by doing Turing computation. Is this a leap of faith?

Evidence indicates that the brain includes mechanisms that resemble digital computation. The pioneering work of Warren McCulloch and Walter Pitts in the 1940s showed that neurons operate discretely, with distinct and identifiable firings that have a binary nature. Either a firing occurs or it does not. McCulloch and Pitts argued that the behavior of any network of neurons could be exactly replicated by a very different network. They argued that functions of the neurons could be described in a propositional logic, and therefore any realization of the same logic would perform the same function that the neurons perform. If this is true, then, in theory, it should be possible to convey my consciousness to some physical machine other than my brain. It is debatable, however, that the logic in the neuron firings completely constitutes consciousness. McCulloch and Pitts' model, for example, assumes that the timing of the firings is irrelevant to the function they perform. This notion seems unlikely.

In the 1960s, the philosopher Hilary Putnam developed the idea that different structures could realize the same function, calling the principle "multiple realizability." Bickle (2016) describes it this way:

> In the philosophy of mind, the multiple realizability thesis contends that a single mental kind (property, state, event) can be realized by many distinct physical kinds.

Under this principle, mental states are not so much dependent on the hardware (the brain) in which they occur, in that other realizations of the same states would realize the same function. In other words, mental states are like software. Again, it is a stretch to conclude that these same states can be realized in a computer. This would require either that computers be universal information-processing machines or that the brain be limited to the same class of functions that computers can realize.

It may seem that the thesis of multiple realizability is reinforced by the distinctly digital encoding in DNA. DNA uses a base-four encoding rather than binary, but it is still digital. A DNA molecule consists of a pair of strands of nucleotides, where each nucleotide consists of one of four nucleobases. The digital genetic code is used to synthesize each new human, and that human realizes cognition. Does this mean that cognition is digitally encoded?

Your offspring may have eyes the color of yours, but they have entirely their own mind. As pointed out by George Dyson in *Turing's Cathedral*, "the problem of self-reproduction is fundamentally a problem of communication, over a noisy channel, from one generation to the next" (Dyson, 2012, p. 287). Recall Shannon's channel capacity theorem from section 7.4, equation (4), which states that a noisy channel can only communicate a finite number of bits of information per use of the channel. Given this limitation, there is no point in encoding more than a finite number of bits in the genetic material. The information would not get through the noisy channel anyway. There would be no point in a nondigital DNA.

Only features that can be encoded with a finite number of bits can be passed from generation to generation, according to the channel capacity theorem. If the mind, or features of the mind such as knowledge, wisdom, and our sense of self, cannot be encoded with a finite number of bits, then these features cannot be inherited by our offspring. It certainly appears that DNA does not encode the mind because the mind of your offspring is not your own or even a combination of those of both biological parents. An infant does not emerge with a fully developed mind. The mind emerges later.

If the mind requires mechanisms beyond digital for its operation and character, then the mind cannot be conveyed by *any* mechanism over a noisy channel. Your mind is entirely your own. Not only can it not be passed on to your offspring, it cannot be passed to anything. It will never reside in other hardware unless we invent a noiseless channel. Biological inheritance cannot provide a noiseless channel because if it did, there would be no mutation, there would be no evolution, there would be no humans, and we would have no minds at all. Genetic inheritance is, of necessity, digital, but minds are formed from more than genetics.

Although DNA is digital and encodes how to construct a brain, a mind does not arise from a brain alone, unless you take an untenably extreme position on the classic nature versus nurture debate. The formation of the mind is heavily influenced by the environment in which it forms, by language, by culture, and by education. Although the brain incorporates some binary operations, like computers, if timing matters, then even the binary reactions are not purely algorithmic, proceeding as a *sequence* of discrete step-by-step state changes, as in a computer. Finally, we just don't know enough about neurophysiological reactions to conclude that they are all purely binary and algorithmic. Consider the effects of drugs, noise, nutrition, and so on on the mind.

John Daugman, a computer science professor specializing in computer vision and pattern recognition at the University of Cambridge, documents a long history of technological metaphors for the brain dating back to the ancient Greeks:

> Theorizing about brain and mind has been especially susceptible to sporadic reformulation in terms of the technological experience of the day. (Daugman, 2001)

He talks about "Freud's hydraulic construction of the unconscious"; clockwork metaphors in Descartes, Hobbes, and many other thinkers; and steam engine metaphors from various writers. The history, he says, is a "stumbling progression toward its inevitable culmination in today's understanding that the brain turns out to be a computer."

Daugman then critiques researchers who "ask precisely that we not think of computation as just the contemporary metaphor, but instead that we adopt it as the literal description of brain function." This converts the "enlivening effect of a new metaphor" into "the deadening effect of embracing one too literally or too ideologically or too long." He concludes,

> While the computational metaphor often seems to have the status of an established fact, it should be regarded as a hypothetical, and historical, conjecture about the brain.
>
> ...
>
> Today's embrace of the computational metaphor in the cognitive and neural sciences is so widespread and automatic that it begins to appear less like an innovative leap than like a bandwagon phenomenon, of the sort often observed in the sociology and history of science. There is a tendency to rephrase every assertion about mind or brains in computational terms, even if it strains the vocabulary or requires the suspension of disbelief.

David Deutsch, a strong proponent of digital physics, also believes that software, as constructed today, will not achieve cognition, which he calls "artificial general intelligence" (AGI):

> [AGI] cannot be programmed by any of the techniques that suffice for writing any other type of program. Nor can it be achieved merely by improving [the] performance [of programs] at tasks that they currently do perform, no matter by how much. (Deutsch, 2012)

Deutsch cites Watson, the IBM computer that defeated former Jeopardy champions Brad Rutter and Ken Jennings in 2011, stating that Watson was not "mimicking human thought processes." He points out that "no Jeopardy answer will ever be published in a journal of new discoveries," whereas, in principle, Rutter and Jennings are both capable of coming up with answers that can be so published. Deutsch also reaffirms my observation that cognition involves processes that are not externally observable, stating, "the relevant attributes of an AGI program do not consist only of the relationships between its inputs and outputs."

But Deutsch is *not* saying that AGI is unachievable with computation. Rather, he is saying that we don't know *how* to achieve it using computation. In fact, Deutsch claims to have *proved* that AGI is, in principle, achievable by Church-Turing computation. His proof relies on the "quantum theory of computation" (more physics that is difficult to understand) to show that everything in the physical world "can, in principle, be emulated in *arbitrarily fine detail* by some program on a general-purpose computer" (emphasis added).

But if you assert that emulating something in "arbitrarily fine detail" is equivalent to actually achieving that something, then you would have to also assert that no meaningful difference exists between rational numbers and a continuum, despite Cantor's observation that there are vastly more real numbers than rational numbers. Given a randomly chosen real number, the probability that it is a rational number is zero, but you can arbitrarily closely approximate it with a rational number.[5] I believe a continuum is qualitatively different from a countable set. In chapter 10, I will give simple examples of real systems where emulation in "arbitrarily fine detail" fails completely to capture essential features of the system.

Penrose, in *The Emperor's New Mind*, goes much further to argue that consciousness cannot be explained by the known laws of physics. Penrose argues that the mind is not algorithmic and must be somehow exploiting hitherto not-understood properties of quantum physics. I am making a less radical argument, in that I argue that the mind is not digital and algorithmic even if it *can* be fully explained using the known laws of physics.

Alan Turing postulated what is known as the "Turing test" for determining whether a computer program realizes a cognitive function. In this test, a human evaluator would observe natural-language conversations between another human and a computer that is programmed to generate human-like responses. The evaluator would be aware that one of the two partners is a computer but would not know which one. Turing said that if the evaluator cannot reliably tell the computer from the human, then the computer is said to have passed the test.

In chapter 11, I will examine what passing the test can tell us broadly about the capabilities of computers. For now, it is evident that if consciousness is not externally observable, then the Turing test tells us nothing about whether a computer (or even a human) has consciousness.

The same questions arise with other brain functions, such as love, empathy, and understanding. Searle put forth a famous argument called the "Chinese room

---

5 See chapter 11 for a discussion of what it means for something to have probability zero.

argument": that no machine operating like a computer, following algorithmic step-by-step rules, can understand natural language.

The Chinese room argument goes like this. Assume that you know no Chinese, either written or spoken (I believe this is true of Searle, at least). You are locked in a room with a small window and a rule book. Someone outside the room, who does understand Chinese, hands you a stack of cards with Chinese characters on them. The cards tell a story in Chinese and then ask a question about the story. You pick up the first card and find in your rule book a matching symbol and a rule telling you what to do with the card. For example, you might put the card in a particular place on the table. You also have a supply of cards within the room, and occasionally the rule book will tell you to find a card in that supply and put it into a pile that you will then pass out through the window. You similarly go through all the cards that were given to you until the rule book tells you that you are done. You then pass the pile you constructed out the window.

Searle then asks a simple question: Did you understand the story? It's quite possible that the answer you gave to the question about the story would lead the external observer to conclude that in fact you did understand the story. However, Searle points out that there is no way you could come to that conclusion yourself.

In this scenario, you are acting exactly like a computer. The algorithmic computation that computers (and Turing machines) perform has exactly this nature. If you are the computer, then no matter how good your rule book is, and no matter how convincing your answers are, you have not achieved what we call "understanding." We can even extend the thought experiment with an unbounded supply of paper with preprinted symbols, giving us a computer with unbounded memory, and the conclusion would not change.

Searle's argument created quite a firestorm of controversy. The community of researchers in artificial intelligence (AI) was not happy. Many counterarguments and counter-counterarguments ensued, some of them really quite entertaining. Google it.

My argument is different, but it leads me to believe that Searle's *conclusion* is probably right, even if you do not believe his argument. There are so many more functions than computations that it is unlikely any given function found in nature is actually performing Turing computations. There just aren't enough possible computations.

This is not to say that software cannot accomplish a great deal. Some people, such as the American computer scientist and futurist Ray Kurzweil, have predicted a "singularity," a point at which a runaway effect takes over, where computers surpass the ability of humans to control or understand what they do. Kurzweil may be right. I hope not, but nothing that I say makes this impossible.

But I *am* saying that achieving *human-like* intelligence, what I call *digital psyche* and what Searle calls *strong AI*, using computers as we know them today is extremely improbable. I believe that we will never consider computers to be siblings, and we will never be able to upload our souls to them even if their capabilities eventually exceed those of the human brain in any or all dimensions. Of course, future forms of software and hardware may have entirely new properties, so I will not make any claims about what they can or cannot accomplish.

9.4 Symbiotic Partnership

The idea of a human-like digital psyche, like digital physics, is a paradigm shift. Perhaps this one too is just waiting for the skeptics, like myself, to die out. Actually, I believe that if anything, it *underestimates* what we can and will do with computers. Forming them in our own image may have biblical appeal, but it is not likely to take full advantage of their complementary capabilities.

I've already mentioned that Google makes me smarter. So does Wikipedia. Sergey Brin, cofounder of Google, once said, "We want Google to be the third half of your brain" (Saint, 2010). What if what is really happening is a coevolution of man and machine, where each becomes more fit for survival or more likely to procreate because of the other?

Google and Wikipedia both do things that no human can do. They exist as data and software in vast server farms in the cloud, as we saw in chapter 5. They have features of a life form, including, for example, the immune system in Wikipedia, where, like a lymphocyte, ClueBot NG kills vandalism, as we saw in chapter 1. Even the Internet can repair itself, with its ability to route around damage. These machines have features of a nervous system, the "dreaming" that is indexing, organizing, and machine learning. Are we playing God, creating a new life form in our own image, or are we being played by a Darwinian evolution of a symbiotic new species? Are humans the purveyors of the "noisy channel" of mutation, facilitating sex between software beings by recombining and mutating programs into new ones, as described in chapter 5?

George Dyson, in *Turing's Cathedral*, raises this question of coevolution. He talks about Google's million-plus servers as a "collective, metazoan organism." He points out that "the companies and individuals who nurture [these servers] are ever more richly rewarded in return" and "unemployment is pandemic among those not working on behalf of the machines." However, it is not just the machines enslaving the humans. The humans are evolving too. "Facebook defines who we are, Amazon defines what we want, and Google defines what we think" (Dyson, 2012, pp.308, 325).

There is no question in my mind that humans are coevolving with computers. If computers and software form organisms, then they depend on us for their procreation. We provide the husbandry and serve as midwives. In exchange, we depend on them to manage our systems of finance, commerce, and transportation. More interesting, the machines make the humans more effective at the husbandry that spreads the software species. Elaborate software simulation of semiconductor physics leads to smaller and less power-hungry transistors. Computer-aided design software enables humans to design billion-transistor chips. Compilers translate human-readable code into machine-readable bits. Google makes it easier for humans to fix problems with the machines (just search for the error message), turning us into their own healing agent. And software innovations fuel the startup culture of Silicon Valley, where the software survives and evolves only if the company survives and evolves, and vice versa.

But it's not just humans making machines more effective. The machines are also making the humans more effective and survivable. Cars are learning to refuse to crash. Hearing aids, pacemakers, and insulin pumps compensate for failures in our bodies. Credit card companies' computers block fraudulent transactions, compensating for failures in our society. Data mining is starting to be able to detect the spreading of diseases, such as SARS and the zika virus. And Google makes it easier for your doctor to find cases that manifest the same combination of symptoms in a human pathology, turning computers into agents in our own healing. As Dyson observes, "[t]he Big Computer [is] doing everything in its power to make life as comfortable as possible for its human symbionts" (Dyson, 2012, p. 313)

Dyson goes further, raising a question that we cannot ignore:

> Are we using digital computers to sequence, store, and better replicate our own genetic code, thereby optimizing human beings, or are digital computers optimizing our genetic code—and our way of thinking—so that we can better assist in replicating them? (Dyson, 2012, p. 311)

Here Dyson is asking about a purposefulness in computers. However, coevolution does not require teleology in nature, so why should it require it in this case?

Coevolution is happening, where computers and humans are both getting more capable. As with much symbiosis in nature, the partnership can empower both partners, even if the genetic manipulations that Dyson asks about do not occur. Wikipedia still makes me smarter, even if I am stuck with the DNA I was born with.

Doomsayers worry that humans will become unnecessary and the machines will dispose of or enslave us. That does not necessarily happen with symbiosis in nature, so why should it happen here? The fungus in lichen has not killed off the algae, even after millions of years. Instead, stronger connections and interdependencies between man and machine could create a more robust ecosystem, such as the notion of an interconnected "nature" that Andrea Wulf says was invented by Alexander von Humboldt (chapter 2).

Evolution is a natural process. It is pointless to simply fear it, and if we understand what is happening, we can help guide it in desirable directions. Creating human-like digital psyches is, in my view, not a desirable direction. Fortunately, it is probably not even achievable, at least not with today's computer designs. Instead, the real power in the partnership between man and machine comes from their complementarity.

To understand that complementarity, we have to understand the fundamental strengths and limitations of both partners. Software is restricted to a formal, discrete, and algorithmic world. Humans connect to that world through the notion of semantics, where we assign meaning to bits. In the next section, I examine the fundamental limits of the world of software and how a partnership between humans and computers can overcome these limitations.

9.5 Incompleteness

Software is limited to a countable, algorithmic world. Humans are not so limited and through the notion of semantics can leverage the countable world of software in an uncountable variety of ways. Limits still exist, however. Scientific thinking strives for rigor, with solid, provable foundations. Mathematics provides the structure for rigorous scientific thinking. However, it turns out that the very quest for rigor results in the same sort of incompleteness that Turing found in computation.

Kurt Gödel published his famous incompleteness theorems in 1931 when he was only 25 years old. His theorems put an end to a decades-long effort known as Hilbert's Program, after the German mathematician David Hilbert. Hilbert's Program, put forth by Hilbert around the turn of the twentieth century, was to put mathematics on a sound foundation as a formal language.

Stephen Hawking, who played a major role in the development of the Bekenstein bound and the digital physics agenda, in a wonderful lecture delivered through a speech synthesizer in 2002 (Hawking, 2002), cites Gödel's theorems. He claims that these theorems do more than just end Hilbert's Program, which pertains to mathematics; they may also end the positivist agenda in science, where every physical theory is a Platonic truth waiting to be discovered. He explains the positivist philosophy as follows:

> In the standard positivist approach to the philosophy of science, physical theories live rent free in a Platonic heaven of ideal mathematical models. (Hawking, 2002)

He makes the connection to Gödel as follows:

> What is the relation between Gödel's theorem and whether we can formulate the theory of the universe in terms of a finite number of principles? One connection is obvious. According to the positivist philosophy of science, a physical theory is a mathematical model. So if there are mathematical results that cannot be proved, there are physical problems that cannot be predicted.

As we will see, Gödel's theorems tell us that within any (consistent) formal system, some statements cannot be proven true or false. So Hawking is saying that, given some formalism for modeling the physical world, inevitably some statements within that formalism we cannot know to be true or false. Although this could be a huge disappointment to scientists striving for that ultimate goal, the grand unified theory, Hawking draws a more optimistic conclusion:

> Some people will be very disappointed if there is not an ultimate theory that can be formulated as a finite number of principles. I used to belong to that camp, but I have changed my mind. I'm now glad that our search for understanding will never come to an end, and that we will always have the challenge of new discovery. Without it, we would stagnate.

With this conclusion, Hawking reaffirms my observation that scientists, like engineers, will never be finished. Although each formalism that we might come up with has its limitations, there is no end to the suite of possible formalisms. There will always be room for invention of new theories.

To understand Gödel's results, we need to first understand what Hawking is referring to as a formulation based on a "finite number of principles." This depends on the idea of a formal language. Before explaining exactly what a formal language is, let me give a simple example, a language that I call X. As with any written natural language, sentences in a formal language are written down as a sequence of characters from an alphabet. X has a small alphabet, with just one letter, x. So the sentences expressible in this language are any sequence of xs, such as "$xxxx$." We can't say much in language X. If I had written this book in language X, then it would be boring indeed.

A formal language has a set of axioms, which are sentences that are by definition true in the language. X has just one axiom, which asserts that the sentence "xx" is true. In a formal language, you cannot argue with the axioms. They are true by definition. So it doesn't really matter whether you *believe* me when I say "xx." When I say "xx" in the language X, it is true. Don't argue.

A formal language has a set of inference rules, which can transform one or more true sentences into another true sentence. X has just one inference rule, which is that if some sentence S is true, then the new sentence Sx (just append an x to S) is also true. For example, if "xxx" is true, then so is "$xxxx$."

What are the true sentences in X? Well, that's an easy question. The true sentences are "xx," "xxx," "$xxxx$' and so on. The only sentence not known to be true is "x," and perhaps the empty sentence, if that is included in the language.

Summarizing what we have so far, a formal language has an alphabet for forming sentences, a (preferably small) set of axioms, and a (preferably small) set of inference rules. That's all. More interesting formal languages will have a bigger alphabet, which might, for example, include the characters $+$ and $\times$ to represent addition and multiplication of numbers. The inference rules could include, for example, basic rules of logic, such as, "If you know that at least one of sentences

A and B is true, and you know that A is false, then you can conclude that B is true." Almost all of mathematics is expressible within such formal languages, including mathematics that deals directly with continuums.

A formal language has a notion of a *proof*. A proof is a sequence of true sentences that demonstrate that a particular end sentence is true. The sequence starts with the axioms, which are assumed to be true, and follows with sentences constructed using the inference rules. The final sentence in the sequence is the sentence proved by the proof. Hilbert's Program sought a formal language that would prove all mathematical truths and have no contradictions (i.e., cannot prove any false sentences).

More precisely, a formal language is *complete* if every sentence in the language can be proved true or false. The language is *consistent* if no sentence can be proved both true and false. Hilbert sought a complete and consistent formal language for mathematics.

In X, every sentence with at least two xs has a proof. For example, the proof of the sentence "$xxxx$;' is the sequence of sentences ("xx", "xxx", "$xxxx$"). This is a proof because it starts with an axiom, ends with the sentence being proved, and uses inference rules to get from one sentence to the next. There is no way in X to construct a proof that a sentence with at least two xs is false, so X is consistent. Is it complete?

The sentence "x" has no proof, but there is also no proof that it is false. In fact, within the language X, I cannot make the sentence, "The sentence 'x' is false." The only sentences I can make are rather boring sentences, such as "$xxxxxxx$." Therefore, this language cannot possibly have a proof that "x" is false. Thus, X is not complete unless I extend it with another axiom, that "x" is false or true. Within the language X, the sentence "x" is neither true nor false. I have to step outside the language X to assign a truth value to "x."

What does it mean for a sentence to be "true" if a formal language can have sentences that are neither true nor false? This question has vexed logicians for some time. One possible resolution to this conundrum is to equate the notion of "truth" with the existence of a proof. This is, in fact, a form of logic called "intuitionistic logic," which ironically is not very intuitive. In intuitionistic logic, a sentence is true only if there is a proof that it is true, and it is false only if there is a proof that it is false. Under this logic, in X, the sentence "x" is neither true nor false. Intuitionistic logic rejects the "law of the excluded middle," an axiom of classical logic that asserts that any sentence must be either true or false. It replaces this law with a constructive principle, which states that truth or falsehood are consequences of constructive demonstrations of that truth or falsehood.

Intuitionistic logic is a rather draconian solution to this problem. A more pragmatic approach is to simply accept that we will sometimes have to rely on certain sensible or self-evident statements as truths even if we have no proof for them. In other words, we may need to assume some elements of the Platonic Good.

Alfred Tarski, a Polish mathematician who later emigrated to the United States, showed in 1936 that no formal language rich enough to possibly satisfy Hilbert's Program could completely define its own notion of truth. In effect, to talk about the "truth" of some sentences in a formal language, you may have to step outside the formal language and use what Tarski called a "metalanguage." This "undefinability of truth" is credited to Tarski, but really it was already present in Gödel's own results, so it might be more appropriate to credit Gödel for it. Nevertheless, Tarski made it explicit and widely known and understood, as much as such a concept can be understood.

Consider now the following sentence, which I will call "Gödel's sentence":

"There is no proof for this sentence."

Suppose this sentence can be made in some formal language (clearly X is not a rich enough language for this). If the sentence is true, then the language cannot be complete because there is at least one sentence, Gödel's sentence, that is true but has no proof. If the sentence is false, then the language cannot be consistent because if the sentence is false, then it has a proof, which means there is a proof for a false sentence. Hence, any formal language that can express Gödel's sentence cannot be both complete and consistent.

Note that I don't have to establish whether Gödel's sentence is true or false. That would require me to use a metalanguage, per Tarski. If the sentence is true, then the language is incomplete. If the sentence is false, then the language is inconsistent. Either way, I've shown that no language that can express Gödel's sentence can satisfy Hilbert's Program.

You've probably already noticed an obvious resolution to the conundrum raised by Gödel's sentence. Let's just avoid any language that can express Gödel's sentence! Are there such languages? Sure. X is such a language. But X can't satisfy Hilbert's Program because it can't express much math. Notice that it can express *a little* math. For example, I can interpret the sentence "xxx" to mean the natural number 3. So I can express all the natural numbers in X (I could interpret the empty sentence, which says nothing, as zero). But the language gives me no way to express, say, "$xxx + xx = xxxxx$" because the symbols $+$ and $=$ are not in the alphabet. The *only* sentences I can make in X are strings of xs.

Hilbert wouldn't be satisfied with X. Is there some other language that would have made him happy? Gödel, to the great disappointment of many optimistic mathematicians, showed that any formal language that is rich enough to describe addition and multiplication of natural numbers can in fact make Gödel's sentence or a sentence logically equivalent to it. Hence, any formal language that has the potential to address Hilbert's Program is either incomplete or inconsistent. This is Gödel's first theorem. Gödel's second theorem, also published in 1931, showed that no rich enough formal language can prove its own consistency. You have to step outside the language and use a metalanguage to construct any such proof.

Gödel's sentence may seem too clever, too cute, a parlor trick. The essence of Gödel's sentence is its self-reference. The sentence talks about itself, reminiscent of the self-scaffolding of software, human consciousness and self-awareness, and Searle's "the concept that names the phenomenon is itself a constituent of the phenomenon." It suggests that formalisms capable of self-reference are all problematic. Hawking points out that such self-reference is also intrinsic in science because the humans who are building models of the physical world are part of that same physical world:

> [W]e and our models are both part of the universe we are describing. Thus a physical theory is self-referencing, like in Gödel's theorem. One might therefore expect it to be either inconsistent or incomplete. The theories we have so far are both inconsistent and incomplete. (Hawking, 2002)

So this is not a parlor trick. It is quite fundamental.

I won't explain how Gödel proved his theorem, although it's an interesting subject. I've already risked losing too many readers. If you are interested in understanding this more deeply, I recommend Franzén (2005), which is informal and accessible. A more rigorous overview can be found in Raatikainen (2015). A delightful and witty exposition of the topic can be found in *Gödel, Escher and Bach: An Eternal Golden Braid* by Douglas Hofstadter (Hofstadter, 1979), which won a Pulitzer Prize.

Instead of giving you more detail on Gödel's theorems, I would like to consider their implications for modeling and software. In Gödel's formal languages, the set of all mathematical statements and the set of all proofs are countable sets, just like the set of all computer programs. Moreover, a "proof" in a formal language is a sequence of transformations of sentences, where each transformation is governed by a set of inference rules. This is conceptually close to what a computer does when it executes a program. In a computer, the sentences in the formal language are ultimately just sequences of bits, and the inference rules are the instructions in an instruction set architecture.

Given a finite alphabet, every sentence constructed as sequences of letters from that alphabet can be encoded as a sequence of bits. The original alphabet and sentence become semantic interpretations of the bit sequences. If each inference rule in a formal language is a computable function, then a proof is exactly an execution of a program. It has to terminate to be a proof, and therefore the existence of a proof and the halting problem are closely connected.

This is not just theory. Extremely useful computer programs, called "theorem provers," take as input an encoding of a sentence in a formal language and attempt to apply the inference rules of the language backward until the program transforms the bit pattern into one or more axioms. If the program succeeds, then the program has constructed a proof. Gödel and Turing both showed, in different ways, that no such program can always succeed.

The formal languages considered by Gödel only permit us to make a countable number of mathematical sentences. Moreover, his incompleteness result only applies to formal languages that are rich enough to describe arithmetic on *natural numbers*, another countable set. If instead we look at formal languages that describe arithmetic on *real* numbers, then the theorems do not apply. Tarski showed in 1948 that a natural theory of real numbers that expresses addition and multiplication, the so-called theory of real closed fields (RCFs), is both complete and consistent (and also decidable, an even stronger property which asserts that the truth or falsehood of any statement can be determined by an effectively computable function). He also showed a theory of Euclidean geometry that is complete, consistent, and decidable.

Formal languages that talk about real numbers are better behaved than formal languages that talk about natural numbers. This notion perhaps further supports my suspicion that Kronecker got it backward when he ascribed the integers to God and the rest to man. It also supports my argument that when designing information-processing machines, we should not limit ourselves to software, which is forced to live within the world of natural numbers.

The number of sentences in any particular formal language is countable, but what about the number of possible formal languages? Given any formal language that fails to provide a proof for some proposition that we care about, we can always come up with a different formal language that does provide such a proof or even one that makes that proposition an axiom. Moreover, even if we restrict the alphabet to, say, zeros and ones, the number of possible semantic interpretations for formal languages is certainly not countable. However, like Kuhn's paradigms, distinct formal languages, particularly those with distinct semantics, are incommensurable. The sentences of one formal language cannot be evaluated through the lens of the other. Making comparisons *across* formal languages requires stepping outside these formal languages into a metalanguage. This is what we do when we assign *semantics* to a syntax. For example, if I interpret the sentence "*xxx*" in the language X to mean the natural number three, then I have assigned it a semantics. The language X has no notion of "three" and assigns no meaning to "*xxx*" except that it is true.

Turing computation is a formal language. The alphabet for this language has two letters, 0 and 1, and the sentences are sequences of bits. The semantics I assign to these bit sequences, however, is quite arbitrary. As I pointed out in section 9.2, I could interpret the bit sequence 11 to mean π, although 11 is not a direct binary encoding of the number π. Hence, just as formal languages can talk about irrational numbers, so can computers, given a suitable semantic interpretation to the bit sequences. The partnership of computers with humans enables such semantic interpretation, and because the number of possible semantic interpretations is much larger than the countable number of formal sentences in any formal language, there is plenty of room for creativity.

Mathematical formal languages on real numbers are widely used by scientists and engineers to specify models. These models can be talking about systems whose behaviors involve uncountable sets, although the total number of sentences that can be formed in any formal language is countable. For example, Faraday's law, equation (256), specifies voltages and currents in a continuum, but the law can nevertheless be encoded in a formal language. In fact, we routinely use models that involve uncountable sets whenever we use a continuum in our models, for example, to model time or space. With suitably chosen semantic interpretations, these models can be manipulated by computers. However, the manipulations are meaningless without the semantic interpretation, so the partnership with humans becomes essential.

If the number of possible systems or behaviors in the physical world is uncountable, however, then given any one formal language and any one semantic interpretation, we can expect that most systems are *indescribable*. There just aren't enough sentences to describe more than a tiny subset. If nature is not constrained to give us only describable systems, then we should assume that with high probability any system found in nature will be indescribable by any particular formal language paired with a semantic interpretation.

Humans are not restricted to working only with one formal language and one semantic interpretation. It is true that given any formal language, such as the accepted language of mathematics based on Zermelo-Fraenkel set theory, we cannot construct more than a countable number of sentences. Yet we can invent new languages. In fact, we do this all the time. Every programming language is a formal language, and we keep coming up with new ones. We are free to assign semantic interpretations to sentences in any formal language. Although a computer produces only zeros and ones, we can interpret a string of zeros and ones to represent a sequence of letters from the alphabet forming sentences in a natural language. This book is produced on a computer, and formally the book is a sequence of zeros and ones, but only a human can give it meaning.

If our goal is engineering rather than science, then we do not need for our new languages to describe some preexisting physical system, such as human cognition. Rather we only need the language to be able to describe something useful or even just something beautiful or interesting. And we need the language to be implementable. We need to be able to find a physical realization that is reasonably faithful to the language. It need not be perfect. No computer is perfect.

Models are similar. They are expressed in a formal language. Speaking as an engineer, I need models to be understandable, and I need to be able to find physical systems that are reasonably faithful to my models. I do not need my models to be true. For example, I know we can make a pretty good balloon machine even if the circumference will never be exactly $\pi \times d$, and we can make a pretty good inductor even if Faraday's law does not exactly describe its behavior.

Mathematical models are capable of describing behaviors that software cannot handle numerically, but by assigning semantic interpretations to the formal symbols manipulated by a computer, computers can help humans to use such models. Computers can, for example, sometimes prove theorems and symbolically solve mathematical equations involving real numbers. However, software is fundamentally limited to a countable world, and it is limited to processes that are algorithmic, following step-by-step procedures. If the physical world is not so limited, then there are machines that can perform functions that software cannot. I believe it is extremely unlikely that the physical world is so limited; thus, despite the amazing things we can do with software, we can't do everything, and even what we can do with software often requires a partnership with humans to give it any semantic meaning. Computers are not universal machines.

10 Determinism

··· in which I argue that determinism is a property of models not of the physical world; that determinism is an extremely valuable property, one that has historically delivered considerable payoffs in engineering and science; that deterministic models may not be usefully predictive because of chaos and complexity; that families of deterministic models that embrace both discrete and continuous behaviors are incomplete; and that nondeterministic models, used explicitly and judiciously, play an essential role in engineering.

10.1 Laplace's Demon

The previous three chapters show definitively that we cannot know everything. Of course, each of us already knew that about ourselves, but these results are more fundamental. They assert that not everything is knowable. If we limit our study to computational processes, those given algorithmically by software, then Turing's result shows that we cannot tell what some programs will do just by looking at the code. If we broaden our study to formal mathematical models, then Gödel's result shows that we will always be able to construct models that we cannot determine to be true or false (or, worse, that end up being both true and false). Moreover, cardinality arguments show that there are many more possible information-processing functions than there are computer programs, mathematical models, or even descriptions in any given language. Consequently, some functions are not computations, cannot be modeled mathematically, and cannot even be described (completely) in any language we have. Unless nature, for some inexplicable reason, limits itself to only the tiny subset of functions that are computable and describable in a language we have, nature will inevitably continue to throw things at us that we cannot understand. To a scientist, this may be frustrating, but to an engineer, it simply means that the horizon for creativity is infinitely far away. There are no bounds to what we can do because we can continue to invent new languages and formalisms, and because they will never be complete, we will never be finished.

Because we can't know everything, we need systematic ways to deal with uncertainty. In the next chapter, I will directly address how to model uncertainty with probabilities. Before we can do that, however, we need to address the question of whether uncertainty is caused by the limits of what we can know or some intrinsic randomness in the world or in our models of the world. Many of the mathematical models and computer programs that we construct are deterministic, which suggests that we should be able to know quite a bit about them unless they fall into the Turing and Gödel traps. Most computer programmers strive to avoid these traps, yielding understandable programs, and also to create deterministic programs. However, this notion of determinism is not a simple one. We can't confront uncertainty without first confronting determinism.

Determinism is a deceptively simple idea that has vexed thinkers for a long time. Broadly, determinism in the physical world is the principle that everything that happens is inevitable, preordained by some earlier state of the universe or by some deity. For centuries, philosophers have debated the implications of this principle, particularly insofar as it undermines the notion of free will. If the world is deterministic, then presumably we cannot be held individually accountable for our actions because they are preordained.[1]

Determinism is quite a subtle concept, as is the notion of free will. John Earman, in his *Primer on Determinism*, admits defeat in getting a handle on the concept:

> This is already enough to make strong the suspicion that a real understanding of determinism cannot be achieved without simultaneously constructing a comprehensive philosophy of science. Since I have no such comprehensive view to offer, I approach the task I have set myself with humility. And also with the cowardly resolve to issue disclaimers whenever the going gets too rough. But even in a cowardly approach, determinism wins our unceasing admiration in forcing to the surface many of the more important and intriguing issues in the length and breadth of the philosophy of science. (Earman, 1986, p. 21)

But Earman insists that "determinism is a doctrine about the nature of the world." I will circumvent the most treacherous difficulties by instead adopting a principle that I first learned from the Austrian computer scientist Hermann Kopetz, who asserted that determinism is a property of models and not a property of the physical world. This thesis does not diminish my fascination with the deep questions that Earman addresses, but it certainly does make it easier to apply the concept of determinism to engineered systems.

As a property of models, determinism is relatively easy to define:

> A model is deterministic if given an initial *state* of the model, and given all the *inputs* that are provided to the model, the model defines exactly one possible *behavior*.

In other words, a model is deterministic if it is not possible for it to react in two or more ways to the same conditions. Only one reaction is possible. In this definition, I have italicized words that must be defined within the modeling paradigm to complete the definition, specifically, "state," "input," and "behavior."

---

1 Author and neuroscientist Sam Harris, in his short 2012 book *Free Will*, argues that even without determinism, free will does not exist. His argument is quite compelling but off topic for this book.

For example, if the state of a particle is its position $x(t)$ in a Euclidean space at a time t, where both time and space are continuums, and if the input $F(t)$ is a force applied to the particle at each instant t, and the behavior is the motion of the particle through space, then Newton's second law, equation (4096), is a deterministic model.

Two useful variations of this idea are evident immediately. First, a model may not have any inputs, in which case it is called a "closed model." For example, if we assume that the universe is everything there is, then any model of the universe cannot have any inputs. Nothing outside the universe exists to provide those inputs. A closed model is deterministic if given an initial state, exactly one behavior is possible.

Second, a deterministic model may be reversible. In this case, given the state of the model at any particular time, and given the inputs at all times (if there are any inputs), both the past and future of the model are uniquely defined. There is only one possible past and only one possible future. Put another way, in a closed reversible deterministic model, the behavior for all time is determined by the state at any one time.

One reason that this simple concept has been so problematic is that all too often, when speaking of determinism, the speaker is confusing the map for the territory. To even speak of determinism, we must define "input," "state," and "behavior." How can we define these things for an actual physical system? Any way we define them requires constructing a model. Hence, an assertion about determinism will actually be an assertion about the model not about the thing being modeled. Only a model can be unambiguously deterministic, which underscores Earman's struggle to pin down the concept.

Consider that any given physical system inevitably has more than one valid model. For example, a particle to which we are applying a force exhibits deterministic motion under Newton's second law but not under quantum mechanics, where the position of the particle will be given probabilistically. However, under quantum mechanics, the evolution of the particle's wave function is deterministic, following the Schrödinger equation (I will say more about this later). If the "state" and "behavior" of our model are the wave function, then the model is deterministic. If instead the state and behavior are the particle's position, then the model is nondeterministic. It makes no sense to assign determinism as a property to the particle. It is a property of the model.

If I have a deterministic model that is faithful to some physical system, then this model *may* have a particularly valuable property: the model may predict how the system will evolve in time in reaction to some input stimulus. This predictive power of a deterministic model is a key reason to seek deterministic models.

We already know that some deterministic models are not predictable. For example, Turing showed that we cannot predict, for all computer programs, whether the program execution will halt for a particular input, even if the program is deterministic. It turns out that there are many more deterministic models that are also not predictable because of chaos and complexity.

In chapter 2, I made a distinction between the engineering and scientific uses of models. An engineer seeks a physical system to match a model, whereas a scientist seeks a model to match a physical system. For these two uses, determinism plays different roles. For an engineer, the determinism of a model is useful because it facilitates building confidence *in the model*. In chapter 4, I talked about logic gates as deterministic models of electrons sloshing around in silicon. The determinism of the logic gate model is valuable: it enables circuit designers to use Boolean algebra to build confidence in circuit designs that have billions of transistors. The model predicts behaviors *perfectly*, in that an engineer can determine how a logic gate model will react to any particular input, given any initial state.

Of course, the usefulness of the logic gate model also depends on our ability to build silicon structures that are extremely faithful to the model. We have learned to control the sloshing of electrons in silicon so that, with high confidence, a circuit will emulate the logic gate model billions of times per second and operate without error for years.

Herein lies an essential difference between science and engineering. To a scientist, for a deterministic model to be useful, it must faithfully describe the behavior of a given physical system. To an engineer, for a deterministic model to be useful, it must be possible to *construct* a physical system that is faithful to the model. In both cases, "faithful" means that the behaviors of the model and the physical system match to a high degree of accuracy. However, the goals are different, and therefore deterministic models will be useful to an engineer in situations where the same models are not useful to a scientist and vice versa.

Some of the most valuable engineering models are deterministic. In addition to logic gates, we also have digital machines, instruction set architectures, and programming languages, most of which are deterministic models. The Turing machines in chapter 8 are also deterministic. The determinism of all these models has proved extremely valuable historically. The information technology revolution is built on the determinism of these models.

For a scientist, fundamentally, when considering the use of deterministic models, it matters quite a lot whether the physical system being modeled is also deterministic. Is the sloshing of electrons in silicon deterministic? If only a model can be unambiguously deterministic, then how can we answer this question? The fact is that almost all established laws of physics are deterministic models, and most are also reversible. Ohm's law for resistors and Faraday's law for inductors from chapter 2 are both reversible deterministic models. For a resistor, if I define the input to be a voltage and the output to be the current, then Ohm's law gives me a deterministic model of a resistor. The model is described by equation (1024). Newton's laws of motion and Einstein's theories of relativity are deterministic. Interestingly, even basic laws used to study the thermodynamics of gasses such as Boyle's law and Charles' law are deterministic, although they do not require the underlying motion of gas molecules to be deterministic. They define state, input, and output in terms of pressure, temperature, and volume not in terms of positions and momentums of the gas molecules. Even quantum mechanics is almost entirely deterministic, in that the evolution of the wave function as defined by the Schrödinger equation is deterministic.

The question of whether the physical world is deterministic has remained unanswered for a long time. In the early 1800s, the French scientist Pierre-Simon Laplace made an argument for determinism in the universe. Laplace argued that if someone (a demon) were to know the precise location and velocity of every particle in the universe, then the past and future locations and velocities for each particle would be completely determined and could be calculated from the laws of classical mechanics (Laplace, 1901). Is this true?

As I've pointed out before, the laws of classical mechanics, such as Newton's second law, equation (4096), are wrong. They need to be adjusted using Einstein's relativity to be precise, although the imprecision will be insignificant for most applications of classical mechanics. Moreover, the notions of position and velocity that underlie the notion of "state" in classical mechanics are undermined by quantum mechanics, although again only significantly at extremely small scales. If the question is the fundamental scientific question of whether the world is deterministic, then *any* imprecision, no matter how small, matters.

What about the probabilistic nature of the wave function in quantum mechanics? Does this undermine the idea of a deterministic universe? Stephen Hawking argues that it does not:

> At first, it seemed that these hopes for a complete determinism would be dashed by the discovery early in the 20th century that events like the decay of radioactive atoms seemed to take place at random. It was as if God was playing dice, in Einstein's phrase. But science snatched victory from the jaws of defeat by moving the goal posts and redefining what is meant by a complete knowledge of the universe. (Hawking, 2002)

Hawking is referring to the fact that the Schrödinger equation, which describes how a wave function evolves in time, is deterministic. "In quantum theory, it turns out one doesn't need to know both the positions and the velocities [of the particles]." It is enough to know how the wave function evolves in time.

Although I can't possibly explain it fully (I'm not convinced that anyone can), it is worth a brief aside on wave functions because they represent the only established nondeterminism that I know of in widely accepted models of fundamental physics. In quantum mechanics, the position of a particle in space is not described simply as a point in a three-dimensional Euclidean geometry but rather by a wave function, a squiggle that changes shape and moves in space over time. The square of the value of the wave function at a point in space and time represents the relative probability of finding the particle at that position at that time.[2] The use of probabilities, a subtle concept that I discuss in chapter 11, implies that the position of the particle is nondeterministic, and the wave function gives the relative likelihoods that an observer will find the particle at any particular point in space.

In 1926, Erwin Schrödinger, a Nobel Prize-winning Austrian physicist, published what is now a key centerpiece of quantum mechanics, the Schrödinger equation. This equation describes the evolution in time and space of the wave function, and as Hawking points out, that evolution is deterministic.

The wave function represents probabilities, and its interpretation is fraught with difficulties. In what is now called the Copenhagen interpretation, originally proposed in the years 1925 to 1927 by the Danish physicist Niels Bohr and the German physicist Werner Heisenberg, the state of a system continues to be defined by probabilities until an external observer observes the state, and only at that point do the probabilities influence the outcome. Prior to being observed, all possible outcomes represented by the probabilities continue to remain possible. This requires an "observer" who is somehow separate from the system and measures the position of the particle. The usual interpretation of a probability is that it specifies the likelihood of observing a particular outcome of an experiment. Under this interpretation, a probability makes sense only if the experiment is performed and the outcome is observed. How can any observer be separate from the physical system?

Schrödinger pointed out difficulties of the Copenhagen interpretation, famously illustrating them with what is now called "Schrödinger's cat." In this thought experiment, a cat is locked in a chamber containing a mechanism that will release a poison if a particular radioactive atom decays, emitting radiation. The decay of a radioactive atom is governed by a wave function, so the decay event is governed by probabilities. The probabilities evolve deterministically according to the Schrödinger equation, but under the Copenhagen interpretation, no actual experiment governed by those probabilities occurs until an observer observes the system. Because all possibilities remain possible until an observer observes the system, Schrödinger argued that the cat must be both alive and dead until such observation occurs. Only then does it become one or the other.

These difficulties have led to endless debate about the meaning of the wave function, with a variety of sometimes bizarre positions emerging. Some of these positions posit that the observer somehow lives outside of physics, that the observer is God, and that this observer is the essence of human cognition. In the 1950s, the physicist Hugh Everett III dispensed with the observer, bundling observer and observed under a single wave function that evolves deterministically under the Schrödinger equation. This view is often called the "many worlds" view because it can be interpreted to mean

---

2 For the position of a particle, this is actually a probability density not a probability. The probability of finding the particle at any specific point in space and time is zero (see chapter 11).

that all outcomes exist simultaneously in an infinite number of parallel universes. In this view, we can take the state of a system to be its wave function, and then quantum dynamics is deterministic.

So Laplace's question still stands, except that now we have to update it to consider the "state" of the system to be represented by its wave function not by the positions and velocities of its particles, and we have to account for the curvature of space time no matter how small. If we make these adjustments, is the resulting model of the universe deterministic? The question of whether the physical world *itself* is deterministic is probably unanswerable.[3] However, we *can* answer the question of whether any particular *model* of the universe is deterministic. We need to keep distinct the map and the territory.

In 2008, David Wolpert used Georg Cantor's diagonalization technique to prove that Laplace's demon cannot exist (Wolpert, 2008). His proof relies on the observation that such a demon, were it to exist, would have to exist in the very physical world that it predicts. This results in a self-referentiality that yields contradictions, not unlike Turing's undecidability discussed in chapter 8 and Gödel's incompleteness theorems discussed in chapter 9. In fact, the result is a kind of essential incompleteness that must result from any deterministic model of the universe, similar to Hawking's observation quoted earlier. Binder (2008), in a review of Wolpert's work, observes poignantly, "It is possible, though, that these various theories, along with all that we have learned in physics and other scientific disciplines, will yet merge into the best science can do: a theory of almost everything." That theory, incomplete as it is, today consists almost entirely of deterministic models.

These debates are fascinating and more philosophical than scientific, but they are largely irrelevant to the engineering use of models. The value of a deterministic logic gate model does not depend at all on whether the sloshing of electrons in silicon is deterministic. It depends only on whether we can build silicon structures that emulate the model with high confidence. We do not need and cannot achieve perfection. As Box and Draper say, all models are wrong, but some models are useful, and logic gates have proved extremely useful.

Although determinism can help predict how a system will evolve in time, I will show in section 10.2 that even a deterministic model may not predict future behavior well. It may be foiled by a phenomenon called chaos; by complexity, where it simply becomes impractical to compute the predictions; or even more simply by an accumulation of error. In such cases, nondeterministic models may become valuable.

---

3 For a concise summary of the various interpretations that have been put forth, see Hoefer (2016). For a more in-depth study of determinism in physical models, Earman's *Primer on Determinism* remains a good analysis of determinism in various physical theories (Earman, 1986).

Every model has a bounded regime of validity. Newton's laws are accurate at modest speeds and scales, but even a system that is well described by Newton's laws may evolve to be outside the regime of validity of the model. Suppose that we apply a modest constant force for a long time to an object in space. Then Newton's second law predicts that the velocity will grow without bound, eventually exceeding the speed of light, in violation of Einstein's special theory of relativity. Compared with an actual physical system, the error in the model's prediction will become arbitrarily large. Nevertheless, for a reasonably short time horizon, with small forces and large masses, Newton's second law will predict behavior extremely well, and such prediction is valuable. Similarly, Einstein's general theory of relativity explains gravity at large scales, whereas quantum mechanics explains interactions of matter at small scales, and attempts to unify these remain unsatisfactory.

Deterministic models may also be foiled by complexity. A classic example from thermodynamics is the pressure exhibited by a gas in a chamber. This can be modeled as collisions of individual molecules with each other and with the walls of the chamber. In Laplace's day, these collisions would have been governed by Newton's deterministic laws of motion, but such models are intractable. Any attempt to compute the individual motions of even a relatively small number of molecules under such laws of motion would overwhelm even the most powerful computers today. As a consequence, physicists consider these motions to be nondeterministic and rely on the statistics of large numbers of random events to exhibit behaviors that are well modeled deterministically by Boyle's and Charles' laws. The emergence of deterministic models from large numbers of nondeterministic behaviors is a consequence of the *law of large numbers*, considered in the next chapter. Our ability to model transistors as deterministic switches relies on similar statistical arguments.

Complex behaviors can arise from even simple models, which can exhibit a phenomenon called "chaos." Interestingly, the inability to make predictions for chaotic models despite their determinism can actually be a valuable property. The technology of encryption, which obscures the content of messages, depends on both determinism (to ensure that the message can be decrypted by the intended recipient) *and* an inability to make predictions (to protect the message from eavesdroppers). Interestingly, although cryptographers today depend on deterministic models, they hope the physical world is actually nondeterministic and some form of "true randomness" can be tapped to make stronger encryption techniques. True randomness, it turns out, is extremely difficult to achieve.

A nondeterministic model may also lend itself to prediction, but instead of predicting a single behavior, it predicts a family of behaviors. Each behavior in the family is possible. A nondeterministic model of a coin toss, for example, simply states that both heads and tails are possible. A deterministic model of a coin toss remains elusive. Even if we laboriously construct one using some exact model of the shape and material properties of the coin and the surface on which it lands, the predictive value of the model would be poor because even the smallest error in our model could drastically change the outcome of a toss. Despite its uselessness, Karl Popper, high priest of scientific positivism, insists on such a model for the toss of a die:

> One sometimes hears it said that the movements of the planets obey strict laws, whilst the fall of a die is fortuitous, or subject to chance. In my view the difference lies in the fact that we have so far been able to predict the movement of the planets successfully, but not the individual results of throwing dice. In order to deduce predictions one needs laws and initial conditions; if no suitable laws are available or if the initial conditions cannot be ascertained, the scientific way of predicting breaks down. In throwing dice, what we lack is, clearly, sufficient knowledge of initial conditions. With sufficiently precise measurements of initial conditions it would be possible to make predictions in this case also. (Popper, 1959, p. 198)

The root of this insistence is a firm belief in the determinism of the underlying physical system. However, predictions can only be made with models, and no such model will be faithful to the physical system in any useful way, so an irreconcilable gap remains here.

Regardless of whether the underlying physical world is deterministic, a nondeterministic model may be augmented with probabilities, which attach a measure of our uncertainty about the outcome. The unfair coin toss considered in chapter 7, where we expect only 1 of 10 tosses to produce tails, can be modeled with probability 0.1 for tails and 0.9 for heads. I will explore probabilistic models in chapter 11, but for now let's just focus on deterministic models.

10.2 The Butterfly Effect

Studying atmospheric effects with the goal of being able to predict weather better, Edward Norton Lorenz came to a disheartening conclusion that prediction beyond a few days was simply not possible, despite that his models were deterministic. Working as a research meteorologist at MIT in the early 1960s, Lorenz was among the first to use well-developed mathematical models of convection and thermal effects in fluids to build computer simulations of weather. He noticed, however, that his models would yield radically different behaviors if he started the simulations with minutely different initial states.

> [T]wo states differing by imperceptible amounts may eventually evolve into two considerably different states. If, then, there is any error whatever in observing the present state—and in any real system such errors seem inevitable—an acceptable prediction of an instantaneous state in the distant future may well be impossible. (Lorenz, 1963)

People later called this extreme sensitivity to initial conditions "the butterfly effect" after a metaphor put forth in the title of a talk by Lorenz, where the turbulence created in the air by the wing of a butterfly could cause a tornado. The butterfly wing changed the initial conditions just enough to make the difference between the tornado forming and the tornado not forming. The tornado would not have formed had the butterfly not flown.

Since the time of Lorenz's initial experiments, computers, mathematical models, and data gathered about weather have all improved by many orders of magnitude. Yet it is still true that predictions beyond about 14 days of weather patterns like rain and wind are not reliable. Indistinguishable initial conditions can lead to radically different weather.

A common feature of models that have such extreme sensitivity to initial conditions is that their behavior can appear random, capricious even. For this reason, the phenomenon is often called "chaos," although the models are actually deterministic. Models of fluid flow, as occurs, for example, as air moves around on earth making weather, frequently exhibit chaos. These models can capture the general pattern of behavior of a system but not the details. The turbulence that you feel in an airplane, for example, has the character of highly random motion (see figure 10.1). Even the most detailed model will not be able to meaningfully predict that motion.

Even simple models can exhibit extreme sensitivity to initial conditions. Figure 10.2 shows the trajectory of a billiard ball on a table with a fixed circular obstacle in the middle. In this case, a small variation in the angle of the initial path of the ball eventually results in a completely different trajectory around the table. Although the starting trajectory of the ball on the left is almost imperceptibly different between the solid line and the dotted line, a radically different trajectory results.

Lorenz's studies of chaos all involved physical systems operating in a continuum of space and time. It turns out that purely digital systems can also exhibit chaotic behavior. The electrical engineer Solomon Wolf Golomb, whom we encountered in chapter 2 for his famous quote, "You will never strike oil by drilling through the map" (Golomb, 1971), figured out that surprisingly simple digital logic circuits could generate bit patterns that appeared to be random (Golomb, 1967).

I first learned about Golomb's "linear feedback shift registers" in the early 1980s when I was at Bell Labs designing modems, devices to transmit bit sequences over an ordinary telephone channel that had been designed to carry human voice signals not bit sequences. It turns out that modems behave much better on seemingly random bit sequences, where there are no repeating patterns. Golomb had figured a way to make almost any bit pattern look completely random using a simple logic circuit called a "scrambler." The original bit sequence can be easily extracted using a similar logic circuit called a "descrambler" at the receiving end. I was so impressed with the elegant

Figure 10.1
Turbulence in a vortex from the tip of an airplane wing tracked with the help of colored smoke. [Image by NASA Langley Research Center, released into the public domain.]

simplicity (and the Boolean algebra that could be used to analyze these circuits) that I made an oil painting with a scrambler circuit and LED lights embedded in the canvas (see figure 10.3). The LED lights exhibit a seemingly random pattern that repeats itself only every 14 hours. The circuit operates reliably to this day.

Pseudorandom patterns were also used in a much more serious art work shown in figure 10.4, a light sculpture by the American artist Leo Villareal. The sculpture consists of 25,000 LED lights installed in 2013 on the San Franscisco Bay Bridge. The lights are controlled by a computer to create patterns that were designed to never repeat during the entire intended two-year lifetime of the installation.

Golomb's circuits generate pseudorandom bit sequences. They seem random but are not. They produce digital chaos. Pseudorandom bit sequences are central to simulation, computer games, cryptography, and even some art. In computer games, for example, they create an illusion of random things happening, whereas in fact the game is completely deterministic.

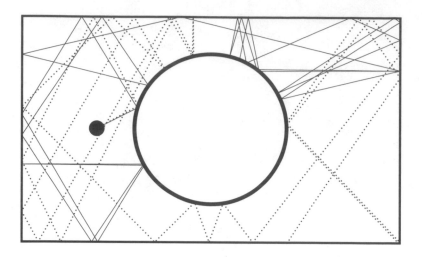

Figure 10.2

A billiard table with a fixed circular obstacle in the middle.

Figure 10.3

Pseudo Random, oil, TTL circuits, and LEDs on canvas, 1981, by the author.

Figure 10.4
The Bay Lights, light sculpture by Leo Villareal (2013).

In the early 1980s, Stephen Wolfram noticed a connection between Golomb's circuits and cellular automata. Cellular automata are simple digital models with a rectangular grid of bits, where each bit gets repeatedly updated by computing some logic function of the neighboring bits. A famous example of a cellular automaton is Conway's Game of Life, developed in 1970 by the British mathematician John Horton Conway. Conway's game is an astonishingly simple deterministic closed model that exhibits tantalizingly lifelike behavior. It captured the imagination of many people, including Wolfram, who devoted much of the rest of his career to studying cellular automata and related phenomena.

The game has a rectangular grid of cells that are either alive (shown as black squares) or dead (white squares). An initial state has some cells alive and some dead, as shown in figure 10.5. At each step of the game, the cells are updated according to the following rules:

1. Any live cell with fewer than two live neighbors dies.

2. Any live cell with two or three live neighbors lives on to the next step.

3. Any live cell with more than three live neighbors dies.

4. Any dead cell with exactly three live neighbors becomes a live cell.

Conway metaphorically associated these rules with life, where underpopulation, overpopulation, and reproduction could all change the state of a cell. Despite the simple rules, the game exhibits surprisingly complex behavior. As the game proceeds, patterns may become stable, like the Block and Beehive shown in the figure, or they may move across the grid, as in the Spaceship. They can also oscillate between two repeating patterns, and they can exhibit seemingly random, chaotic

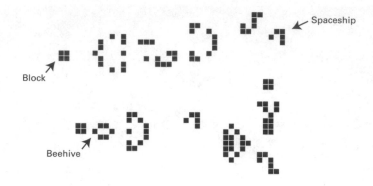

Figure 10.5
A snapshot of Conway's Game of Life.

behavior for quite a long time. It is mesmerizing to watch one of these games play out.

Conway's game is purely digital, easily realized on a computer. The fact that such simple rules can exhibit such complex behavior inspired Wolfram, who in his 2002 book, *A New Kind of Science*, concludes that "all is computation" (Wolfram, 2002). More specifically, Wolfram postulates that all natural processes can be constructed out of simple rules, and the complexities arise because of the chaos that such rules can induce. He makes a compelling and engaging case, but of course the ultimate "truth" of such a postulate would depend on digital physics.

Despite chaos, many engineered systems are predictable with high confidence with time horizons of years. Transistors are good examples. Although any detailed model of the underlying motion of electrons in the silicon will be chaotic, the macroscopic behavior of a transistor is simple. It functions as a switch. A gasoline engine in a car is another example. The explosions in the cylinders are highly chaotic, but they reliably deliver controllable power to the powertrain. Harnessing chaos is a key goal of engineering and, if Wolfram is right, of nature as well.

10.3   Incompleteness of Determinism

Laplace believed that nature can be fully described by deterministic models. Wolfram goes further and argues that nature behaves like *computational* models that are also deterministic. The set of deterministic computational models is much smaller than the set of deterministic models. The set of computational models excludes continuums, for example. So Wolfram's claim is more aggressive than Laplace's.

In both cases, the models may exhibit chaos, so they are capable of describing immensely complex behavior. But the chaos also limits the utility of the models as predictors, making Laplace's demon a difficult concept to accept. Nevertheless, the models are deterministic.

In Laplace's world, time and space are continuums through which objects move. In Wolfram's world, time and space are discrete grids of cells that are updated in a step-by-step fashion. What happens if we assume that the world has both kinds of behaviors, discrete and continuous? I will give a simple example that suggests that in such a world, determinism is incomplete. Specifically, a set of deterministic models that describes the world using a mixture of discrete and continuous behaviors has "holes" in it, situations that should be able to be modeled deterministically but cannot be. To patch these holes, we either have to disallow discrete behaviors altogether, asserting that they do not occur in the physical world, or we have to embrace digital physics and sacrifice almost all known physical models, including relativity and quantum mechanics, both of which model space and time as continuums.

As an example of a model with both continuous and discrete behaviors, consider the collisions of billiard balls, as shown in figure 10.6. Suppose that the left ball is moving toward the right ball with momentum P and the right ball is sitting still, as illustrated in figure 10.6(a). Assume that the surface is frictionless, so the momentum of each ball remains constant until a collision occurs. As long as no collision occurs, the behavior is continuous.

Suppose that we model a collision as a discrete event. That is, we assume that the collision occurs in an instant, having no duration in time. Such a model needs to determine the momentum of the balls after the collision as a function of their momentums before the collision.

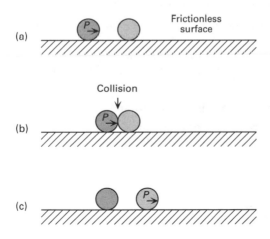

Figure 10.6
Collision of ideal billiard balls on a frictionless surface.

Assume that the balls are ideally *elastic*, meaning that no kinetic energy is lost when they collide. In this case, Newton's laws require that both energy and momentum be conserved, in the sense that the total momentum and energy in the balls must be the same after the collision as before.[4] In the physical world, some of the kinetic energy will be converted to heat due to friction, but here we will assume that doesn't happen or that so little kinetic energy is lost that we can neglect it.

Assuming the masses are the same, there are two possible results from the collision that preserve both momentum and energy. One result is that the left ball passes right through the right ball without interacting with it. This outcome could happen, for example, if the ball were actually a neutrino rather than a billiard ball. However, for billiard balls, this outcome is extremely unlikely, so we are justified in rejecting this possibility. The only other result that conserves both momentum and energy is that the balls exchange momentums, as shown in figure 10.6(c). The left ball is now still, and the right ball is moving away at the same velocity that the left ball was approaching before the collision.

Now suppose that the two balls have different masses. It turns out that in this case, there are still exactly two possible solutions, one where the left ball passes through the right ball and the other where they bounce. Let's again choose the more reasonable solution where they bounce. Because the masses are different, after the collision, *both* balls will be moving. If the left ball is heavier than the right, then they will both be moving to the right. If the left ball is lighter than the right, then they will be moving in opposite directions. In both cases, their speeds after the collision are uniquely determined by Newton's requirement that both momentum and energy be conserved. Hence, the model is deterministic.

If there are more than two balls, however, then the situation gets much more interesting. Consider a thought experiment where two billiard balls are approaching a third ball from opposite sides, as illustrated here[5]:

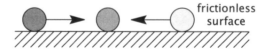

Assume that the center ball is sitting still and the two outer balls collide with the center ball at the same time. To keep things simple, let's start with the assumption that all three balls have the same mass. What will happen?

---

4 The momentum of a ball is the product of its velocity and its mass. The energy of a ball is half of the product of the mass and the *square* of the velocity.

5 The details of this thought experiment are given in Lee (2016). I will spare you the nerd storm here, but if you want to check the conclusions, please see that article. Penrose (1989) also used multiball collisions to show that the notion of determinism even in classical mechanics is problematic. His examples are a bit more complicated because they occur in more dimensions.

I hope you have enough practical experience with billiard balls that your intuition matches mine. I would expect in this situation that the two outer balls will bounce off the center one and move away from it at the same speed that they had been approaching it. Thus, after the collision, the situation will look like this:

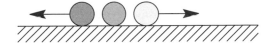

But coming up with a discrete model that predicts this behavior turns out to not be so easy.

A first attempt will be to simply use the same technique that we used with two balls, where the balls exchange momentums when they collide. However, if the collisions are simultaneous, then the left and right balls will exchange momentums with the center ball at the same time, and these momentums will have opposite signs, canceling each other out. Thus, all three balls will suddenly stop. This solution fails to conserve both momentum and energy.

An alternative way to handle this situation is to treat the two simultaneous collisions as a sequence of two-ball collisions with no time elapsing between the collisions. As shown in figure 10.7(b), when the collisions occur, we can arbitrarily pick one of the collisions to handle first, ignoring the other collision. Suppose we handle the left collision first, ignoring the right collision, as indicated in the figure. The left ball transfers its momentum P to the middle ball and stops. Without time elapsing, we find ourself in state (c) in the figure, where the middle and right balls are traveling toward one another and colliding. Now there is only one collision, so we handle it in (d), leaving us in state (e), where the balls have exchanged momentums. Again, without time elapsing, a new collision occurs, which we handle in (f), leaving us in state (g). After time elapses, we find ourselves in state (h), where the left and right balls are moving away from the center ball, which has not moved and remains still. This behavior is the one we expected intuitively, where the two balls are moving away at equal speeds after the collision.

In this solution, if the masses of the balls are all the same, then it does not matter which of the two collisions we handle first. Here comes the rub. If the masses of the balls are *not* the same, then the solutions are *not* the same. If we handle the left collision first, then we get one solution. If we handle the right collision first, then we get a *different* solution, with all three balls moving at different speeds.

Suppose, for example, that the center ball weighs twice as much as the left and right balls. To be concrete, let's suppose that the center ball weighs two kilograms and the outer balls weigh one kilogram each (these billiard balls are quite heavy, but nice round numbers make the math easier). Suppose that the left and right balls are approaching the center ball at one meter per second so that they collide simultaneously with the center ball. First, notice that this scenario is completely symmetric, and the same intuitive solution works, where the two outside balls bounce off the center one so that after the collision, they are moving away from the center ball at one meter per second, and the center ball remains still. This solution conserves both momentum and energy.

However, this is *not* the solution we get using the strategy shown in figure 10.7. In that strategy, we handle the left collision first, and then without time elapsing, we handle the second and third collisions that result. I will spare you the nerd storm, but after the sequence of collisions in the figure, the left ball will be moving to the left at about 0.48 meters per second, the middle ball will also be moving to the left, but more slowly, at about 0.37 m/s, and the right ball will be moving to the right at about 1.22 m/s. If you do the math, you can verify that with this solution, both momentum and energy are conserved.[6] Notice that with this solution, the situation is no longer symmetric, although the starting state was symmetric.

So what happens if we handle the right collision first? In that case, we will end up with the mirror image asymmetric solution, where the middle ball is moving to the right. This solution also conserves momentum and energy.

We now have a real conundrum. We have three possible outcomes: a symmetric one derived intuitively and two mirror-image asymmetric ones derived using the strategy of figure 10.7. Newton's laws give us no basis for preferring any one of these solutions. All three solutions, and many more, are consistent with Newton's laws. They all conserve momentum and energy. Because there is more than one allowed behavior, Newton's laws (with discrete collisions) result in a nondeterministic model.

How do we know which behavior will match some physical experiment? Here's where things get tricky. To conduct such an experiment, we will have to ensure that the collisions are actually simultaneous. This will be difficult to do (impossible, in fact, given quantum mechanical uncertainty principles). First, suppose that the collisions are not actually simultaneous but are just close in time. That is, one of the two outer balls collides with the center ball just before the other outer ball arrives. In this case, instead of a single collision among three balls, a sequence of collisions occurs between two balls. This makes the problem easier to solve because when only two balls are colliding, only one outcome after the collision is possible that conserves both momentum and energy (barring the tunneling outcome, where the balls pass through one another). Consequently, if the collisions are not simultaneous, the model remains deterministic. Only one final behavior is allowed by the model.

If the masses of the balls are different, then the behavior is different if the left ball collides first than if the right ball collides first. This will be true no matter how small

---

6 The total momentum in the system after these collisions is $-0.48 \times 1 - 0.37 \times 2 + 1.22 \times 1 = 0$, same as the starting momentum. The total energy in the system after the collisions is $((-0.48)^2 \times 1 + (-0.37)^2 \times 2 + (1.22)^2 \times 1)/2 = 1$, same as the starting energy.

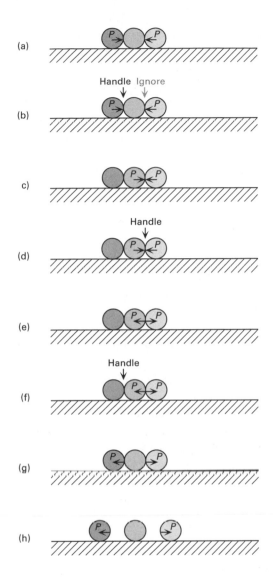

Figure 10.7
One of two orderings for handling collisions among three balls.

the time is between collisions. Let's call the time of the left collision t_L and the time of the right collision t_R. Let the time difference be $d = t_L - t_R$. Consider a sequence of experiments where d is always positive (the right ball always collides first), and d approaches zero, getting smaller and smaller. As d gets close to zero, there will be little difference between the outcomes of these experiments. Changing d from, say, one nanosecond to 0.5 nanoseconds will not change the outcome much. Eventually, as d gets small, there is no significant difference between the experimental outcomes, so the sequence of experiments seem to be converging to behavior that should be the behavior when $d = 0$.[7] It would seem that the limiting behavior should be the one unique behavior when the collisions are simultaneous.

However, it is not. If we repeat the same sequence of experiments, but this time we require d to always be negative, then again we get a sequence of behaviors that are closer and closer together, and they appear to be converging to a behavior, but they do not converge to the same behavior as the previous sequence of experiments.

In the limiting case, when d reaches zero, the collisions become simultaneous. At this point, the behavior will depend on which sequence of experiments we are conducting, the one where $d > 0$ or the one where $d < 0$. These two sequences of experiments converge to different behaviors as d approaches zero. When the collisions become simultaneous, we have no basis for choosing between these two possible limiting cases, so we have to assume they are both possible. They both conserve momentum and energy.

In the single unique experiment where d is exactly zero, there is more than one possible outcome from the model, so the model is nondeterministic. However, if d is not zero, no matter how small it is, then the model is deterministic. A mathematician would call the set of all these deterministic models *incomplete* because this set does not contain its own limit points. In the limit, when d exactly hits zero, the model becomes nondeterministic. The set of deterministic models has a hole at exactly the point where $d = 0$.

Note that it will not do to just disallow $d = 0$ because to do so we would have to disallow t_1 and t_2 to vary smoothly. Assuming time is a continuum, as nearly all current models of physics do, t_1 can cross t_2, in which case there is a point where $t_1 = t_2$, and hence $d = 0$. Note that this point is *even harder* to avoid if we quantize time, as required by digital physics. However, we don't want to do that anyway because doing so sacrifices almost all of modern physics, including Newton's laws, the Schrödinger equation, and Einstein's relativity.

I argued in section 9.3 that a close approximation of a process does not have the same properties as the process unless no significant difference is found between a continuum and a countable set. This billiard ball thought experiment reinforces that arbitrarily close approximations can in fact be quite different from the real thing. If the "real thing" is simultaneous collisions, then this thought experiment shows that all arbitrarily close approximations to it are deterministic, but the real thing is not. Moreover, we can make two scenarios, one with $d < 0$ and one with $d > 0$, that are arbitrarily close to one another in all their parameters but that exhibit wildly different

---

7 Technically, such a sequence of experiments, where the difference between them becomes vanishingly small, is called a Cauchy sequence, after the French mathematician Augustin-Louis Cauchy. A space of models is said to be *complete* if every Cauchy sequence in the space converges to a model in the space. This space of deterministic models is incomplete, as proved in Lee (2016).

(but still deterministic) behaviors. We are forced to conclude that coming arbitrarily close, in "arbitrarily fine detail" in the words of Deutsch, does *not* achieve the "real thing." In this billiard ball experiment, all close approximations of simultaneous collisions are deterministic, but the actual simultaneous collisions are not.

Any model for this physical system that treats the collisions discretely will suffer this problem. What exactly does it mean to treat the collisions discretely? In this case, it means that the model talks about time before and after the collision, but the collision itself does not occupy any time. It occurs instantaneously. An instant before the collision, we have a certain energy-momentum arrangement, and an instant after, we have another energy-momentum arrangement, and the model only requires that total energy and momentum are conserved.

An alternative is to not treat the collision as a discrete event. Instead of being instantaneous, it takes time. The collision starts when the molecules of the balls are close enough to begin to affect one another, and it ends when they have moved far enough apart that they no longer significantly affect one another. We can build this kind of model using either classical mechanics (Newton's laws) or quantum mechanics (using the Schrödinger equation) to describe the continuous evolution of a wave function. Both approaches will yield a model that is deterministic but extremely sensitive to initial conditions, and therefore chaotic.

The nondiscrete classical mechanics solution is reasonably easy to understand. Suppose the balls are ever so slightly springy. That is, when the molecules of one ball get close enough to those of the other ball to start interacting, the molecules of the balls get squished together, like a spring being compressed. The balls slow down. The spring compression temporarily converts kinetic energy to potential energy, so energy is still conserved. As the springs that are the balls compress, the motion of the balls slows until the balls stop. The springs then start to decompress, converting the potential energy back to kinetic energy and pushing the balls apart. A model like this is deterministic, but it is extremely sensitive to the initial positions, speeds, and springiness of the balls. If those are slightly off, then a radically different behavior will result.

So we have a choice. We can either accept a discrete model of the collisions, in which case we lose predictability to nondeterminism, or we can reject discreteness, constructing a detailed model of molecular interactions, in which case we lose predictability to chaos. In both cases, we lose predictability. The model that accepts discrete collisions is much simpler than the one that models molecular interactions, so it seems like the simpler model that admits nondeterminism is the better choice.

10.4   The Hard and the Soft of Determinism

Determinism focuses our attention on a single behavior, a single reaction, a single "right answer" to the question of how a system will react to a stimulus. As an intellectual tool, it is valuable. Being able to identify the "right answer" is essential to Popper's principle of falsifiability in science. An experiment falsifies a theory if its behavior deviates by more than measurement error from the right answer predicted by the theory.

In the complementary engineering use of models, a deterministic model defines the "correct behavior" of a system. Any physical system that deviates significantly from that correct behavior is a flawed implementation of the model. A clear definition of correct behavior is a principle that underlies all of digital technology. The very notion of "digital" discretizes the messy physical world, unambiguously differentiating zero and one, yes and no, right and wrong. It is the ultimate of Serres' hard versus soft.

However, we have to carefully avoid the tarpit that results when we conflate the map with the territory. Determinism is a clear and unambiguous property of models but a muddy and treacherous property of physical systems. Nearly all fundamental models in physics are deterministic, and the question remains open whether intrinsically nondeterministic behaviors are found in the physical world. This question is philosophical more than technical because any nondeterministic model of the physical world may simply reflect an unknown unknown, some hidden variable that determines the outcome but remains invisible to us.[8] Because our knowledge of the physical world can never be complete, as observed by Hawking, we can never definitively assert that the physical world is nondeterministic. Hidden variables or unknown laws of physics may later reveal themselves as our technology and ability to measure and observe the physical world improve.

Regardless of whether the physical world is deterministic, humans have learned to build physical devices, transistors, that exhibit behavior that is extraordinarily faithful to a discrete, digital, deterministic model. The transistors in your laptop computer can switch on and off billions of times per second and operate for years without deviating from the correct behavior defined by a deterministic model. Such fidelity to a deterministic model is unrivaled by any other human-created artifact.

Thus, determinism in models is valuable, but this does not in any way imply that engineers should forgo nondeterministic models. In fact, nondeterminism is an essential tool for overcoming the limitations of determinism. For example, if a deterministic model is too complex to analyze, a nondeterministic model may be

---

8 In quantum mechanics, Bell's theorem, named after the Irish physicist John Stewart Bell, uses quantum entanglement to rule out hidden variables as the source for experimentally observed randomness in quantum systems. Specifically, these experiments show that taking a measurement at one point in space can instantly affect the outcome of another experiment at a remote location, seemingly in violation of the speed-of-light limits on communication. Einstein called this property of quantum physics "spooky action at a distance." Interestingly, although Bell's theorem seems to indicate that randomness in the physical world is real, an equally explanatory resolution is that the world is actually deterministic in an extremely strong way, where every particle carries with it since its inception all the outcomes of all possible measurements that might ever be taken on it any time in the future. Hence, even this result does not definitively prove that randomness is real in the world.

more valuable. A nondeterministic model exhibits multiple behaviors, but if *all* of these behaviors are demonstrably acceptable, then a nondeterministic model is just as good for building confidence in a design.

Nondeterminism has also played a central role in the theory and practice of computer science, even though Turing machines and algorithms are usually deterministic. Concurrent software, where multiple programs execute simultaneously, can be extremely difficult or even impossible to model deterministically, for example.

Nondeterminism even plays a central role in one of the key open questions in computer science: whether P equals NP. This question was apparently first raised in a 1956 letter written by Kurt Gödel to John von Neumann. The question is whether it is always true that a problem where the solution is easy to check is also easy to solve. The "N" in "NP" stands for "nondeterministic," but it would take us too far afield to fully explain this question here. If you are interested, Wikipedia has a nice article.

In science, the models of a theory must predict the behavior of a physical system, or else the theory is not falsifiable and therefore not scientific, according to Popper. However, deterministic models do not necessarily yield predictability. Deterministic models are often extremely sensitive to initial conditions, exhibiting chaos. Even purely discrete, digital, and computational deterministic models can exhibit extremely complex chaotic behaviors.

In engineering, in contrast to science, predictability may be less important than repeatability. Repeatability ensures that an engineered system will behave as designed, with high confidence. This is valuable even if the design is able to exhibit behaviors that are too complex to predict.

According to Wolfram, the ability that digital technologies have to exhibit extremely complex behaviors justifies a belief in a "new kind of physics" that models the physical world in a purely discrete, computational way. I have argued in chapter 8 that I find it extremely unlikely that nature limits itself to the small set of processes that are computational, but even if it doesn't, computational models of the physical world are still valuable. Simulation models of weather, for example, are deterministic, chaotic, discrete, and computational, and nevertheless do a remarkable job of predicting weather for at least a few days into the future. Even if their predictive value is limited, these models lend enormous insight into the processes of nature, even if the mechanisms of the model do not match the mechanisms of nature.

Perhaps more disturbing than chaos is the incompleteness of deterministic models. Any modeling framework rich enough to include Newton's laws of motion, if it also admits discrete behaviors, is incomplete. Therefore, the family of deterministic models has "holes," limiting cases that cannot be modeled deterministically. The philosophical implications of this observation seem profound. It implies that models of the physical world that admit both discrete and continuous behaviors must also admit nondeterminism.

Both chaos and nondeterminism limit the predictive value of models. It appears that both chaos and nondeterminism are unavoidable. Because our ability to predict is limited, our vision of the future will always remain uncertain. Any rich enough modeling framework for practical usage is going to have to deal with uncertainty. In the next chapter, I examine how to confront and manage uncertainty.

11 Probability and Possibility

··· in which I consider the meaning of probability, which I argue is fundamentally a model of uncertainty about a system and not directly a model of that system; and in which I reconsider continuums and argue that probabilistic models over continuums reinforce the conclusion that digital physics is extremely unlikely.

11.1 The Bayesians and the Frequentists

Scientists seek models of the physical world. Even if the physical world is actually deterministic, Laplace's agenda to explain the natural world with deterministic models is foiled by complexity, the incompleteness of determinism, the unpredictability of chaos, Wolpert's proof of the impossibility of Laplace's demon, or indeed all four. If we put aside the incompleteness of determinism (by disallowing models with discrete behaviors), then Laplace's agenda may still be alive but only as a philosophical question. It ceases to be a question of science or engineering because either the explainer (the demon) is impossible (per Wolpert) or any explanation has little or no predictive value due to chaos. Models without predictive value cannot be falsified and hence fail Popper's test for being scientific. Add to this Hawking's observation that Gödel's theorem implies that models of nature will never be complete and Turing's result that we cannot tell what some programs will do just by looking at the code, and it seems we have no choice but to accept uncertainty.

Note that we have to accept uncertainty even if we tenaciously cling to Einstein's "God does not play dice." Regardless of whether the physical world has a phenomenon of random chance, our models cannot be certain. Once we have models that embrace uncertainty, those models will be robust to the philosophical question of whether chance exists in the physical world. The models will work whether it does or not.

Engineers, in contrast to scientists, seek physical realizations of models. In the previous chapter, I argued that determinism is a valuable property of models, and hence if we can find physical realizations that are faithful to deterministic models, we get a powerful partnership. However, there are limits to this approach, just as there are limits to the scientific use of deterministic models. In addition to the unpredictability of chaos, we can easily find ourselves in a situation where we just don't know enough about the physical realizations to be able to construct faithful deterministic models. In this situation, it is valuable to be able to explicitly model our lack of knowledge. This is what probabilistic models do.

Embracing uncertainty also gives us a way to manage complexity. Given any intrinsically complex system, such as the Airbus A350 considered in chapter 6, even if we could construct a deterministic model of its behavior, that model would likely be incomprehensible. By embracing uncertainty, we change the goal. Instead of seeking certainty, we seek confidence.

But what *is* uncertainty, and how do we model it? Nondeterministic models give us a simple way to model uncertainty. We simply create models that have more than one allowed behavior. However, this is often too simple. Nondeterministic models, by themselves, give no indication about which of the allowed behaviors is more likely. In fact, they are missing a notion of likelihood altogether.

So now I could play a dirty trick on you. I could define randomness to be uncertainty, uncertainty to be likelihood, likelihood to be probability, and probability to be randomness. I could couch all that in fancy language so you don't even notice that it's circular. I could even throw the words "stochastic" and "measure" to lend gravitas, and you would still not know what I'm talking about. Instead, I'm going to be up front about it. Probability is circularly defined. It is an axiomatic formal system with solid, well-understood mathematical properties. It's only the interpretation, how to apply the model to the physical world, that is difficult.

Probability is a surprisingly controversial theory from a philosophical perspective, despite that it is well established and mature mathematically. One view, which happens to be my view, is that probability is a formal model for quantifying what we don't know. Perhaps the reason for the controversy is that any time we talk about what we don't know, we can't really know what we are talking about.

We can begin by understanding the distinction between nondeterminism, which talks only about *possibilities*, and probability, which attempts to *quantify* uncertainty. How uncertain are we actually? Connor, channeling Michel Serres, points out that the distinction between probability and possibility is arguably one of the many hard versus soft oppositions, as discussed in chapter 4.

> The gap between probability "uncertain but calculable" and possibility "certain but incalculable" is itself graspable only as another inflection of the hard and the soft. (Connor, 2009)

Probabilities quantify known unknowns. The unknown unknowns can only be handled with possibilities. Possibilities are modeled with nondeterminism, but to have a calculable theory of uncertainty, we need probabilities. For example, probability theory enables us to become more certain about something as we gather data while still preserving room for learning from still more data, leaving a residual uncertainty. The method for doing this is credited to Reverend Thomas Bayes, an eighteenth-century English statistician and theologian. Interestingly, the modern mathematical formulation is due to Laplace, who was apparently hedging his bets about whether his demon could actually remove all uncertainty.

I used probabilities in chapter 7, where I talked about fair and unfair coins. In that chapter, the fair coin had a probability 0.5 of heads, whereas the unfair coin had a probability 0.9 of heads. What do these numbers really mean? In answering this question, we run amuck of a long-standing philosophical debate. Although experts in probability theory pretty much agree on the mathematical machinery that they use, they disagree on the basic meaning of these numbers. These experts fall into two camps: the frequentists and the Bayesians. In the Bayesian model, a probability is a quantified estimate of uncertainty. In the frequentist model, a probability is a statement about how repeated experiments are likely to turn out. Let me explain.

Consider the statement "the probability of heads is 0.5." To a frequentist, this means "in repeated tosses of the coin, on average, half of the outcomes will be heads." To a Bayesian, this same statement means "I have no idea whether a toss will yield heads or tails, so I have no reason to expect one outcome over the other." The frequentist statement is about repeated independent experiments, whereas the Bayesian's statement is about uncertainty, about what we don't know.

Consider now the statement about the unfair coin, "the probability of heads is 0.9." To a frequentist this means, "If I repeatedly toss the coin, on average, 9 out of 10 outcomes will be heads." To a Bayesian this same statement means, "I believe strongly that a coin toss will very likely yield heads." The number 0.9 is a quantification of the strength of this belief.

If you toss the unfair coin N times, both frequentists and Bayesians expect to see heads $N \times 0.9$ times if N is large enough. This is an informal statement of a central tenet of probability called the *law of large numbers*. Specifically, this law asserts that for large enough N, the actual number of heads, call it M, will be close to $N \times 0.9$. Here, "will be close to" means that the probability that M/N differs from 0.9 by more than some small number ϵ is very small.[1] Even more specifically, "very small" means that for any particular choice of ϵ, we can make this probability as small as we like by choosing a large enough N.

Despite the agreement between the frequentists and Bayesians about the law of large numbers, deep differences exist. The frequentists take the law of large numbers to be the essential definition of probability. A probability is exactly a statement about repeated experiments, no more, no less. A Bayesian, however, uses a probability as a measure of uncertainty, a subjective concept, and hence can interpret the probability to have meaning even if there is only *one* experiment.

Suppose you are tossing the unfair coin of chapter 7, which has a probability of heads of 0.9. Suppose that your first toss of the unfair coin yields tails. I will be surprised. Suppose now that the second toss again yields tails. I will be even more surprised. Suppose the third toss *again* yields tails. Now I am astonished.

---

1 Nerds like to use the Greek letter ϵ (epsilon) for small differences. This even creeps into the vernacular, where a Nerd may assert, for example, "I am within epsilon of finishing this book."

The frequentists and Bayesians would agree that this sequence of events is extremely unlikely, and they both assign a probability of $0.1 \times 0.1 \times 0.1 = 0.001$ (one in a thousand) to this outcome. Both would say, "Wow, that was *really* unlikely," and they might suspect that something is amiss, but then they would part ways.

Both frequentists and Bayesians use probabilistic models to learn about the world. A frequentist might approach the coin conundrum by designing a scientific experiment to test the hypothesis that the coin you are tossing has a probability of heads of 0.9. He would call this hypothesis the "null hypothesis" and would then postulate an alternative hypothesis, for example, that the coin is actually a fair coin.[2] Then the experiment would begin. You would toss the coin once. Suppose it comes up heads. The frequentist would dispassionately observe that the probability of that occurrence was high, 0.9. Suppose the next toss yields tails. The frequentist would say, "Well, that was unlikely." The probability that a single coin toss yields tails is 0.1, but more interestingly, in two repetitions of the experiment, we saw an outcome of tails. The probability of seeing at least one tail in two coin tosses is $0.1 \times 0.9 + 0.9 \times 0.1 + 0.1 \times 0.1 = 0.19$ (because there are three ways that could have happened).

This probability of 0.19, or 19%, is called a "*p*-value." It is a measure of the likelihood of seeing an observation at least as extreme as what was observed (one tail in two tosses) given the null hypothesis. The probability of 19% is small but not small enough to reject the null hypothesis. It could still be true that the coin is the unfair one.

The experiment would continue. You toss the coin again. Suppose it comes up tails again. We have now seen *two* tails in three coin tosses, which is quite unlikely if the coin is unfair in this way. Examining all the possible ways that three coin tosses could yield at least two tails, we come up with a *p*-value of 0.028, or 2.8%. The frequentist would now say, "Hmm... this *p*-value is below my threshold of 5% for rejecting the null hypothesis. I therefore conclude that the coin you are tossing is not the unfair coin I thought it was."

The frequentist's approach to this problem is objective and scientific, in the sense of Popper. He formulated a hypothesis, designed an experiment to falsify it, and rejected the hypothesis when the data indicated that the hypothesis was likely false.

The Bayesian, however, would approach this same problem in quite a different way. She too would suspect that the coin you have tossed is not the unfair coin she thought it was. But she would start by quantifying her initial uncertainty. Let's suppose she sold you an unfair coin that has probability of heads of 0.9.

---

2 Actually, the alternative hypothesis could be much broader. It could be that the probability of heads is some unknown value less than 0.9. This hypothesis includes even the possibility that the coin is some other unfair coin, for example, one where the probability of heads is 0.1. This would not change the experiment or its conclusions in any way.

She therefore believes that the coin you are tossing is very likely to be that particular unfair coin. She assigns a number to this belief, saying, for example, that she is 80% sure you are tossing the coin she sold you. This is a subjective judgement that she calls a *prior probability*. It is a measure of uncertainty prior to any observation of data.

Armed with this prior probability, the experiment begins. Suppose that your first coin toss comes up heads. The probability that the unfair coin comes up heads is 0.9, so she is not surprised. This seems to reinforce her prior probability. She now uses Bayes' formula, which gives her a specific way to update her prior probability, taking into account the new data.

Let U denote the assertion that the coin is the unfair coin she sold you. The Bayesian's prior probability is $p(U) = 0.8$. Let H denote the outcome that the coin toss yields heads. If the coin is unfair, then the probability of this outcome is written $p(H|U) = 0.9$, which is read "the probability is 0.9 of getting heads given that the coin is unfair." Bayes' formula then gives our Bayesian a way to update her prior probability as follows:

$$p(U|H) = \frac{p(H|U)p(U)}{p(H)}. \tag{2}$$

The left side of this equation, $p(U|H)$, is read "the probability that the coin is the unfair one given that a coin toss yielded heads." This new probability is an updated belief called the *posterior probability*. It quantifies our uncertainty about whether the coin is the unfair one *after* observing data. Bayes' formula, therefore, gives us a systematic way to update our subjective beliefs upon making observations of the world.

We have enough information to calculate the numerator on the right side because we know that $p(H|U) = 0.9$ and $p(U) = 0.8$. The only hard part is the denominator, which is the probability of seeing heads regardless of whether the coin is unfair. This probability is a weighted average of probability of seeing heads if the coin is unfair (0.9) and the probability of seeing heads if the coin is fair (0.5), where the weights are the probability that the coin is unfair (0.8) and the probability that the coin is fair (0.2), respectively. So the denominator works out to $0.9 \times 0.8 + 0.5 \times 0.2 = 0.82$.

Putting it all together, evaluating (2), our Bayesian calculates the posterior probability to be 0.878. She is now 87.8% sure that the coin is unfair. The observation of heads has strengthened her conviction that the coin is unfair. Before observing any data, she was 80% sure. Now she is 87.8% sure.

Let the experiment continue. Suppose that the next coin toss yields tails. This outcome is less likely under the assumption that the coin is unfair, so it should result in a lowered confidence that the coin is unfair. Our Bayesian will again apply Bayes' formula (2). She won't bore us with the details but reports that Bayes' formula gives us a new posterior probability of 0.59. She is now only 59% sure that the coin is unfair. Notice that the outcome of tails diminished her confidence by more than an outcome of heads reinforced her confidence. This is because an outcome of tails is much more unlikely than an outcome of heads, so observing this outcome carries more information. In fact, in chapter 7, I showed how Shannon calculated that observing a tail carries about 22 times as much information as observing heads (3.32 bits vs. 0.15 bits), assuming the unfair coin.[3] Because there is more information in this observation, our Bayesian learns more and adjusts her prior probability more.

Continuing the experiment, suppose that the next coin toss *again* yields tails. Then by Bayes' formula, after observing this, our Bayesian will be only 22% sure that the coin is unfair. She now actually believes it is more likely that the coin is fair than unfair. The frequentist came to the same conclusion and rejected the hypothesis that the coin is unfair. But the frequentist had no mechanism to take into account the prior information that the Bayesian sold you an unfair coin. The frequentist's experiment is therefore more objective, but it also omits information that we actually have.

If we continue the experiment and observe yet another tail, our Bayesian will only be 5% sure that the coin is unfair. She is now quite sure that the coin is not the unfair coin she sold you. But it took more observations to reach that level of confidence because her prejudice had to be overcome by the data. The frequentist had no mechanism for taking into account that prejudice.

The Bayesian approach embraces subjectivity. This is consistent with an interpretation of probability as a measure of uncertainty rather than a measure of a percentage of outcomes of repeated experiments. Uncertainty is necessarily subjective. No objective physical reality can be uncertain about anything. The notion of uncertainty is a human cognitive notion.

The Bayesian interpretation of probability is completely invulnerable to the debate about whether the physical world is actually deterministic. If the world really is deterministic, then the frequentists are on a slippery slope. What does it mean to repeat an experiment? If the starting conditions of each repetition are the same and the world is deterministic, then the outcomes should also be the same. This makes the frequentist's experiment useless. So clearly the starting conditions need to be different. But how are they different? How much are they different? It seems that the variability of initial conditions would be the only source of differing outcomes, but the frequentist ignores this variability. Or is it just that the frequentist is uncertain about those starting conditions? But then, isn't the frequentist also faced with subjective uncertainty?

---

3 In chapter 7, we assumed the coin was the unfair one. We can adjust the information content calculation to take into account that we are only 80% sure that the coin is unfair. With this adjustment, the probability of heads becomes $p(H) = 0.82$ instead of 0.9. The information content in an outcome of heads becomes 0.29 bits instead of 0.15. Correspondingly, the probability of tails becomes $p(T) = 0.18$ instead of 0.1, and the information content in observing tails drops from 3.32 bits to 2.47 bits.

The frequentist interpretation of probability is problematic if you assume a deterministic physical world. The Bayesian interpretation is not. The Bayesian approach uses probability models to quantify uncertainty, and regardless of whether the physical world is deterministic, there is no shortage of uncertainty. Moreover, the Bayesian's approach systematizes *learning*, which is the process of reducing uncertainty. We can learn from data in less ad hoc ways. It's no wonder that the Bayesian approach dominates in the field of machine learning, where computer programs continually update probabilistic models of the world. The vandalism detector of Wikipedia that we saw in chapter 1 is an example of such a machine learning application.

Frequentists and Bayesians use pretty much the same mathematical framework, including Bayes' formula (2). Although their experimental methodology differs for the previous coin toss experiment, often they don't differ at all between the two camps. It is odd to see two major camps of intellectuals who disagree so fundamentally and yet almost always agree. It's like Dr. Seuss's *The Butter Battle Book*, where an arms race breaks out between two camps that differ only on which side the bread should be buttered on. The difference between the Bayesian and the frequentists is more philosophical, but at that level the difference is profound.

The Shannon notion of information dovetails nicely with the Bayesian interpretation. The intuitive notion of information is that it counters ignorance. Receiving information results in updating our model of the world or learning. The more information we receive, the more we update our model. If a toss of the unfair coin yields heads, then we are not surprised. This is what we expected. The information content is low (only 0.15 bits, as calculated in chapter 7).[4] So we don't update the model by much (80% to 87.8% in the earlier example). In contrast, if the toss yields tails, then we are a bit surprised. That was not a likely outcome, according to the probability of 0.1 that we assigned to it. The information content is higher (3.32 bits, as calculated in chapter 7).[5] We update the model by quite a bit more. If we fail to learn something from repeated observations of tails and fail to update our model, then we are just being stubborn and dogmatic.

In fact, Bayesian probability gives us a way to understand dogma. If I am initially *absolutely* sure that your coin is unfair, then my prior probability is $p(U) = 1$, and Bayes' formula provides no possibility for learning. In Bayes' formula (2), we will find that if $p(U) = 1$, then $p(H|U) = p(H)$, so the posterior probability $p(U|H)$ will equal the prior probability $p(U)$. No matter what the outcome of coin tosses, Bayes' formula will not change our minds. Even Reverend Bayes cannot overcome stubborn dogmatism. If you are determined not to learn, then you won't learn, no matter what you observe in the world. Bayes' formula proves this.

---

4 Or 0.29 bits after adjusting the calculation to use the prior.

5 Or 2.47 bits after adjusting to use the prior.

Karl Popper, in *The Logic of Scientific Discovery*, objected strongly to the Bayesian approach (which he called the Laplacean approach). He argued that the Bayesian interpretation of probability is subjective, and subjectivity has no place in science.

> It treats the degree of probability as a measure of the feelings of certainty or uncertainty, of belief or doubt, which may be aroused in us by certain assertions or conjectures. In connection with some nonnumerical statements, the word "probable" may be quite satisfactorily translated in this way; but an interpretation along these lines does not seem to me very satisfactory for numerical probability statements. (Popper, 1959, p. 135)

In other words, such "feelings" should not be assigned numbers. He then credits the English economist John Maynard Keynes, whose 1921 *A Treatise on Probability* refined Laplace's and Bayes' notion by interpreting a probability as a "degree of rational belief." To Popper, this interpretation can more rationally be assigned a number, but it is still subjective. Popper minces no words in preferring the frequentists' approach, which he asserts is objective:

> I declare my faith in an *objective interpretation*; chiefly because I believe that only an objective theory can explain the application of the probability calculus within empirical science. (Popper, 1959, p. 137, emphasis in the original)

He then admits that "the subjective theory ... is faced by fewer logical difficulties than is the objective theory," but that this is because they are "non-empirical ... they are tautologies." It is a theory built on its own assumptions, like an axiomatic theory in mathematics. Popper declares this to be "utterly unacceptable."

The frequentist perspective seems to have the advantage of better testability, appealing to Popper's preference for empirical methods. Specifically, a Bayesian approach always requires a prior probability, which is used without being tested. But instead of confirming or falsifying a hypothesis, a Bayesian will adjust the prior probability based on new evidence. How is this less empirical than the frequentist approach? A frequentist will treat repeated coin tosses as a scientific experiment designed to falsify the hypothesis that the coin is unfair. But when should falsification occur? When should the hypothesis be rejected? Frequentists use an ad hoc measure, where thresholds of 5% or 1% for the p-value are common. How is this less subjective? Even with the unfair coin, a string of tails is certainly possible. It's just not likely.

Bayes' rule provides a systematic way to learn from the observations. The Bayesian approach is consistent with the observation that all models are wrong and provides a way to improve the models. The objective approach only provides a way to reject the model, but as Kuhn points out, because all models are wrong, all hypotheses should be rejected. Nevertheless, even the frequentists aren't so rigorous and fall back on ad-hoc confidence measures such as thresholds for the p-value to determine when to reject a hypothesis.

Laplace distinctly adopted a subjective approach that Popper deems unacceptable, but this is actually a consistent position for Laplace to take. After all, Laplace believed in a deterministic world governed by deterministic models and predictable by his demon. In such a world, repeated experiments are pointless. But Laplace recognized that we don't know the initial conditions for the experiments exactly. We are uncertain about those conditions, and his probabilities model exactly that uncertainty, not some intrinsic chance in the world, where God plays dice.

The eighteenth-century Scottish philosopher David Hume supports Laplace's position (and likely influenced Laplace):

> Though there be no such thing as Chance in the world; our ignorance of the real cause of any event has the same influence on the understanding, and begets a like species of belief or opinion.

> There is certainly a probability, which arises from a superiority of chances on any side; and according as this superiority increases, and surpasses the opposite chances, the probability receives a proportionable increase, and begets still a higher degree of belief or assent to that side, in which we discover the superiority. [in *An Enquiry Concerning Human Understanding*]

Popper's position, unlike Laplace's, emphasizes statistics rather than uncertainty. To Popper, probability is not about what we don't know but rather about aggregate behavior of large numbers of individually deterministic players. This point of view is nicely illustrated by his description of a waterfall:

> Imagine a waterfall. We may discern some odd kind of regularity: the size of the currents composing the fall varies; and from time to time a splash is thrown off from the main stream; yet throughout all such variations a certain regularity is apparent which strongly suggests a statistical effect. Disregarding some unsolved problems of hydrody-namics (concerning the formation of vortices, etc.) we can, in principle, predict the path of any volume of water — say a group of molecules — with any desired degree of precision, if sufficiently precise initial conditions are given. Thus we may assume that it would be possible to foretell of any molecule, far above the waterfall, at which point it will pass over the edge, where it will reach bottom, etc. In this way the path of any number of particles may, in principle, be calculated; and given sufficient initial conditions we should be able, in principle, to deduce any one of the individual statistical fluctuations of the waterfall. (Popper, 1959, p. 202)

Popper accepts Laplace's deterministic world, but he really picked a difficult illustration. Models of fluid flow are notoriously chaotic (see figure 10.1), so Popper's "any desired degree of precision" is not really achievable, no matter how precise the initial conditions are. They would have to be perfect. So here we see starkly the debate. The Bayesian says that we don't know where the molecule will go (whether we *can* know is irrelevant) and uses probability to model that uncertainty. Popper says we can know, but we are interested in the aggregate behavior of many molecules, and we use probability to model the aggregate behavior.

Although both perspectives have merit, I personally find the Bayesian perspective more compelling for three reasons. First, the Bayesian approach embraces the notions of information and learning. In Popper's approach, any description of the aggregate behavior of deterministic molecules is either wrong or right, and it can be tested by observing the molecules in the waterfall. If through observation we find that our aggregate model is wrong, then we can update the model and try again. But that update (for Popper) must occur in a meta theory, outside the theory of probability, because probability does not model what we know and don't know, it just models aggregate behavior. The update of the model becomes subjective and possibly capricious because in principle the theory can't help us perform that update. It is ironic that many machine learning techniques used today, like those used in the vandalism detector of Wikipedia (see chapter 1), use Bayesian models. The irony is that machine learning algorithms are completely mechanized, operating without human intervention, and yet, according to Popper, they are subjective. It is hard to reconcile these observations.

Second, the Bayesian approach makes more sense when talking about rare events. A Bayesian can say something like, "The probability of a major earthquake in San Francisco in the next 30 years is 63%" (Field and Milner, 2008). I don't see how a frequentist could rationally make such a statement. For sure, such a statement is not falsifiable and is not a statement about aggregate behavior of many individually deterministic behaviors.[6] Such a statement, if it is backed up by rigorous research, reflects the aggregate *opinion* of many experts and the use of computer simulation models that are informed by prior experience with real earthquakes. However, there is nowhere near enough such prior experience to adopt a frequentist's interpretation of the probability. I nevertheless find statements about rare events useful (if scary). They quantify what we know and don't know, and reasoning about rare events is an essential part of engineering safety-critical systems, such as the Airbus A350.

Third and finally, I find the Bayesian perspective more compelling because it covers situations that seem to be simply not well handled by the frequentist interpretation.

---

6 Unless, I suppose, we adopt Everett's many worlds interpretation of quantum mechanics, which postulates that many worlds exist in parallel at the same time and in the same space as our own. Hence, there are many San Franciscos, some of which will have earthquakes and some of which will not. However, even that theory does not permit any observation of the many worlds, so it is still not falsifiable.

Suppose, for example, that a coin is flipped, but the outcome of the flip is obscured by a cup before you can observe it. What now is the probability that you will see heads when the cup is removed? No matter how many cup-removal experiments you perform, the outcome will always be the same, so it seems that the frequentist would have to say that the probability of heads is either zero or one, but we don't know which it is. The Bayesian, however, has no difficulty with this situation. The probability of observing heads is the same as it was before the coin was flipped.

11.2 Continuums, Again

So far, I have only considered probabilities of events with a finite number of possible outcomes. A coin toss yields either heads or tails. An earthquake either occurs or not. The real world, however, is often messier. A coin could get stuck in the mud and yield neither heads nor tails. Earthquakes are occurring all the time, although fortunately most of them are too small to feel. Many of the things we don't know have more than a finite number of possible outcomes. Probability theory needs some adaptation to reason about those.

Consider a thought experiment. Suppose you throw a dart and you measure the distance between where you are standing and where the dart lands. This process is physical, and I will define its "output" to be the final distance to the dart. Assume that distance is a continuum (see chapter 9), you can measure the distance precisely, and you measure the distance in inches as a real number. For example, the dart may land 120.5 inches away, which is 10 feet plus half an inch.

Now, what is the likelihood that the distance to the dart is an integer? Intuitively, I hope you see that this is extremely unlikely. In fact, with a reasonable probabilistic model of this process, the probability that the outcome is an integer is zero. There are vastly more noninteger distances than integer distances.

OK, you may say, let's measure the distance in millimeters instead of inches. What is the likelihood now that the distance is an integer? If the measurement is precise, then the probability will again be zero. There are *still* vastly more noninteger distances than integer distances. In fact, the probability of getting an integer will be zero *for any* precision of measurement.

However, things get even weirder. Suppose I throw the dart, and it lands at exactly 120.123 inches. Note that I do not require that we actually be able to measure this distance. The dart had to land somewhere, and some oracle knows that it landed at 120.123 inches, even if I don't know this. In a reasonable probabilistic model, the probability that the dart will land at 120.123 inches is zero. In fact, the probability that it will land at *any* specific distance is zero, and yet it lands somewhere. Wherever it lands has a probability of zero. Don't give up on probability yet. Bear with me.

For problems with a finite number of possible outcomes, such as a coin toss, a probability is a number between zero and one, where zero means that the event will not occur, and one means that it always occurs. With my dart experiment, we can no longer interpret a probability of zero to mean that the event will not occur. Under that interpretation, the dart could not land. It would have to remain suspended in mid air because every point where it could land has a probability of zero, but it does land.

Before I give you the details, let me point out a flaw in my argument already. I have asked you to imagine a physical scenario, that of throwing a dart, and use it to draw conclusions about a model, the model of distance measures in a continuum. I am asking you to confuse the map with the territory.

OK, we are going to have to pick one. Do we want to do this experiment in the physical world or in the world of models? If we do it in the physical world, then we face a number of difficulties. First, I asked you to measure the distance *precisely*. You can't do that in the physical world. Even the LIGO experiment makes imprecise measurements of distance, although through an extraordinary feat of engineering and a $1.1 billion investment, they have reduced the imprecision to much less than the diameter of a proton, but not to zero.

To avoid these difficulties, let's do the experiment in the world of models. Now it is a thought experiment. Now that we are in the world of models, we can assume that the dart has a distance, a real number, even if we can't measure it precisely.

The essential problem here is that distance to the dart has a truly vast number of possible values. In fact, it has an uncountably infinite number of possible values. But some values *are* more likely than others. It is unlikely that the dart will land one mile from me. It is also unlikely that it will land very close, say on my feet. How can we model this variability in likelihoods if all distances have a probability of zero?

In probability theory, the way this situation is handled is using a probability *density* rather than a probability. Specifically, I talk about the probability that the dart will land between 9 and 11 feet from me. Perhaps that probability is 0.68. A frequentist would interpret this to mean that in repeated dart throws, 68% will land between 9 and 11 feet. A Bayesian would interpret this as a belief that it is a bit more likely that the next throw will land between 9 and 11 feet than that it will land outside that range. Either way, I assign a probability to a *range* rather than to a particular number. Every individual number will have a probability of zero, but a range can have a probability bigger than zero.

A probability density function for our dart-throwing experiment is shown in figure 11.1. That figure has a characteristic shape called a "bell curve," which suggests that the dart is most likely to land in the vicinity of 10 feet because that is where the curve is highest. As we get farther away from 10 feet in either direction, the probability density decreases, but it actually never gets to zero. In such a curve, the area under the curve indicates the probability of landing within a range. For example, the area of the shaded region is about 0.68, indicating that there is a probability of 68% of landing between 9 and 11 feet. Notice that the value of the

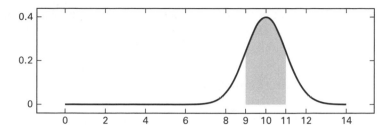

Figure 11.1
Probability density function for the dart distance experiment.

curve at a particular point is *not* the probability that the dart will land at that point. The probability of landing at any one point is zero. The probability of landing in a range is the area under the curve for that range.

Recall from chapter 7 that Shannon's model of information states that a rare event carries more information than a common event. If an event has a probability of zero, for example, the event that the dart lands at 120.123 inches, then an observation of that event carries an infinite amount of information, as measured in bits. Shannon nevertheless uses a *finite* number to model the information content in the dart throw. This finite number is a continuous entropy, which as I explained in section 7.4 does not measure information in bits. Continuous entropy is more like a probability *density* than a probability. We can interpret it as an information density, not as a measure of the information content (in bits) of an individual observation. This allows us to *compare* information densities, just as we can compare likelihoods. If the dart lands one mile from me rather than 10 feet, then it is a truly remarkable event, carrying quite a lot of information, although the probability of both outcomes is zero. Using densities allows me to make such comparisons.

The particular shape of the curve in figure 11.1, the bell curve, is special in probability. Shannon's notion of entropy helps to explain why. The bell-shaped probability density function is called a "normal distribution" (presumably because it is more "normal" than abnormal distributions) or a "Gaussian distribution," after the nineteenth-century German mathematician Carl Friedrich Gauss. As it happens, the normal distribution is the most random of all distributions that have a fixed mean and variance and no other constraints.[7] If you compare the entropy of two random

---

7 The mean, also called the expected value, is the center of mass of the probability density function. For a normal distribution, the mean is where the distribution peaks, or equivalently, the outcome where it is equally likely to get a result above as below it. For a frequentist, the mean is the average value of an infinite number of experimental outcomes. For a Bayesian, the term "expected value" makes more sense than average, as it captures the subjectivity of probabilities. The variance is the average (or expected value) of the square of the outcomes minus the mean. It measures how variable the outcomes are likely to be. If the variance is large, then many outcomes will be far from the mean, whereas if it is small, then most outcomes are close to the mean.

variables with the same mean and variance, one with a normal distribution and the other with some other distribution, the one with the normal distribution has higher entropy and hence is more random.

11.3 Impossibility and Improbability

In the dart thought experiment, when we toss the dart, it lands somewhere. The distance from me to where it lands is a random number selected from a continuum of possible distances. Assuming that distances form a continuum, the probability that any particular distance occurs is zero. However, a particular distance does occur. So in a continuum, a probability of zero does not indicate impossibility but rather indicates extreme improbability.

Now, risking confusing the map for the territory, if we assume that continuums exist in the physical world, then we would have to conclude that any particular thing that happens in the physical world has a probability of zero, at least according to the Bayesian model of probability. Perhaps this is fundamentally why precisely exact measurement of anything in the physical world is not possible. It would imply certainty about something that is extremely improbable. Under the Bayesian interpretation, "probability zero" means that we are *almost* certain that the event in question cannot occur. How could we have confidence in a measurement that reveals something that we are quite sure cannot have occurred? Probability theorists actually use such terminology, where they talk about something that will "almost surely" not occur, by which they mean it has a probability of zero, but it is still possible for it to occur.

Perhaps this is the biggest reason that the Bayesian interpretation of probability appeals to me so much more than the frequentist interpretation. All models are wrong, so certainty is unachievable. Nevertheless, some models are useful, and we use them to build confidence. Models and experiments can narrow our ignorance, but certainty is a fiction. We need to be open to continuing to learn. The Bayesian model gives us a systematic way to do that.

This is also why I do not believe that the digital physics explanation of the physical world is likely to be correct or useful (see chapter 8). Unless we believe that nature limits itself to countable sets, a postulate that is not testable, then the probability that any naturally occurring process is a computation is zero. It is not impossible, but it is extremely improbable. There are vastly many more processes than there are computations.

Human cognition is a naturally occurring physical process. This reasoning, therefore, leads to the conclusion that the probability that human cognition is computation is zero. There are so many fewer computations than physical processes that it is extremely unlikely that nature somehow ended up using only processes from the limited set that we call computation. It is not impossible, but it is extremely improbable.

So we should certainly not just *assume* that cognition is computation, as many people today do. Instead, we need evidence to support the claim. What kind of evidence? How much evidence? Bayes' formula provides guidance because it explains how to update our beliefs based on observations.

But Bayes' formula now presents us with a serious conundrum. Suppose that in equation (2) we let U represent the statement "cognition is achievable using computer programs." Then the above cardinality reasoning seems to require us to assert that the prior probability $p(U)$ is zero. If we take this to be the prior, then no observation H will result in any posterior probability $p(U|H)$ that differs from zero. We are stuck with a dogmatic position that cognition is not computation, and no amount of evidence will change our minds.

I don't like dogma. Under the Bayesian interpretation, a prior probability is subjective. It is not derived from a formula or an experiment. If we have *any* doubt, therefore, we should not set the prior probability to zero. Based on my cardinality argument, however, it seems that our prior should be small. Because there are so many more physical processes than computations, it is unlikely but not impossible that cognition is computation. We might, for example, choose a prior of 1%, $p(U) = 0.01$. This says we believe there is a 1% chance that cognition is computation. What evidence then is required to convince us that cognition is actually computation? What observation H will yield a posterior $p(U|H)$ greater than 50%, convincing us it is more likely that cognition is computation than that it is not?

Bayes' formula (2) states that our belief gets updated when we observe some experimental outcome H by multiplying our prior by[8]

$$E = \frac{p(H|U)}{p(H)}. \tag{1}$$

Let's call this quantity E the "evidence." If E has a value larger than one, then it is evidence in favor of U. If E has a value smaller than one, then it is evidence against U. If E equals one, then it is no evidence at all because our posterior probability will be the same as the prior probability.

In section 9.3, I considered whether the Turing test could help us determine whether a computer program has consciousness. I concluded in that section that if consciousness is not an externally observable property of the brain, then passing the Turing test provides no evidence of consciousness in a program. So if U is the postulate that a program has consciousness, and H is a passing of the Turing test,

---

8 By Messerschmitt's law (see footnote on page 14), it doesn't matter what I write from here on. I am now truly feeling free.

then it must be that $p(H|U)/p(H) = 1$ or equivalently that $p(H|U) = p(H)$. What does this mean?

The quantity $p(H|U)$ is the probability of passing the Turing test given that the program has consciousness, and $p(H)$ is the probability of passing the Turing test regardless of whether the program has consciousness. Thus, for these two quantities to be equal, we just need it to be true that whether a computer program can pass the Turing test does not depend on whether the program has consciousness.

Indeed, many computer programs have been created that quite convincingly pass the Turing test, at least within limited domains. To my knowledge, none of the authors of these programs claims that their program has consciousness.

In a famous example, in the 1960s, the MIT computer scientist Joseph Weizenbaum wrote a computer program called ELIZA that was capable of engaging humans in a conversation. ELIZA would respond to humans in a manner bearing a remarkable resemblance to an empathic psychologist. Using a modern online implementation of ELIZA called "elizabot," I had the following short conversation with ELIZA:[9]

> ELIZA: Please tell me what's been bothering you.
> YOU: I think my computer is smarter than me.
> ELIZA: Do computers worry you?
> YOU: Yes
> ELIZA: You are sure.
> YOU: No, not really.
> ELIZA: Are you saying no just to be negative?

ELIZA was the first example of programs that are now called "chatterbots." Chatterbots are widely used for customer service call centers, airline reservation systems, toys (such as *Hello Barbie*), and automated personal assistants such as Apple's Siri and Amazon's Echo. Malicious chatterbots appear in chat rooms for surreptitious advertising and to entice people to reveal personal information.

Weizenbaum claimed that ELIZA demonstrates that the appearance of human-like intelligence does not in fact imply the existence of human-like intelligence:

> [M]achines are made to behave in wondrous ways, often sufficient to dazzle even the most experienced observer. But once a particular program is unmasked, once its inner workings are explained in language sufficiently plain to induce understanding, its magic crumbles away; it stands revealed as a mere collection of procedures, each quite comprehensible. The observer says to himself "I could have written that." With that thought he moves the program in question from the shelf marked

---

9 http://www.masswerk.at/elizabot/

"intelligent," to that reserved for curios, fit to be discussed only with people less enlightened than he. (Weizenbaum, 1966)

Weizenbaum's goal with ELIZA was to demonstrate that the appearance of intelligence was easy to achieve and therefore should not be construed as evidence of real intelligence:

The object of this paper is to cause just such a reevaluation of the program about to be "explained." Few programs ever needed it more. (Weizenbaum, 1966)

An aspect of Weizenbaum's statement is disturbing. He seems to be claiming that if a program is comprehensible, then it must not be intelligent. This implies that if intelligence is computation, then it must be a form of computation that we can't understand. Doesn't this in effect assert that we will never understand intelligence?

The notion of understanding human intelligence is self-referential because the process doing the understanding must itself be intelligent. This notion therefore is likely vulnerable to the sort of incompleteness that Gödel found in formal languages, Hawking applied to physics, and Wolpert found in Laplace's determinism. If this incompleteness is real and we will never fully understand intelligence, then Weizenbaum's criterion is valid, and we should dismiss intelligence in a program if we fully understand the program. However, I know from experience that it is not hard to write programs that nobody can fully understand. I've written quite a few myself, so this criterion is not really all that limiting. Certainly programs that exhibit digital chaos, as discussed in section 10.2, have inexplicable behaviors.

So what kind of evidence is needed to convince us that cognition is computation? Assuming that our prior is low (e.g. $p(U) = 0.01$), then we need some sort of observation H, where the evidence $E = p(H|U)/p(H)$ is much bigger than one. In other words, we need for $p(H|U)$ to be much bigger than $p(H)$. Hence, we need an observation H that is much more likely if U is true than if U is not true. To get us to a posterior of 50%, where it is equally likely that cognition is computation as that it is not, we need an observation H that is 50 times more likely if U is true than if U is not true. When the prior is small, the burden of proof is higher than if the prior is more moderate.

Alternatively, we can use many observations of weaker evidence. If the Turing test provides any evidence at all, it is at best weak evidence, where $p(H|U)$ is only slightly larger than $p(H)$. Equivalently, passing the Turing test is only slightly more likely if cognition is computation than if it is not. In the face of weak evidence, E is only slightly bigger than one, so if the prior is small, the posterior will still be small. It will take many instances of such weak evidence to overcome the small prior.

One alternative interpretation is that chatterbots *approximate* cognition, and as the software gets more sophisticated, the approximation gets arbitrarily close to actual cognition. Is getting arbitrarily close sufficient? I argued in section 9.3 that we cannot assume a close approximation of a process has the same properties as the process itself unless no significant difference is found between a continuum and a countable set. This observation is reinforced in section 10.3, where a billiard ball thought experiment has the property that all close approximations of simultaneous collisions are deterministic but the actual simultaneous collisions are not. If a close approximation to cognition is not actually cognition, then it is easy to show that the Turing test provides *no evidence* that a program achieves cognition. Under these assumptions, we can, in principle, make a computer program where $p(H)$ is as close as we like to $p(H|U)$. As the program gets more sophisticated, the evidence for cognition E gets arbitrarily close to one, which is the value for no evidence at all.

So things don't look good for the claim that cognition is computation. If the Turing test can provide only weak evidence and the prior is low, then it will take a large number of observations to build support for this claim. If the Turing test provides no evidence, then what other experiment should we conduct to update our low prior probability that cognition is computation? The evidence is not there, so the claim that cognition is computation is ultimately faith based. A claim that cognition is computation amounts to an unreasonably high prior, given the cardinality argument and no effective method for updating the prior with experimental evidence. This is the essence of faith.

I'm an engineer not a scientist. My goal isn't to model the physical world or replicate natural processes such as human cognition. My goal instead is to make physical systems that do interesting and useful things. I believe we can do interesting things with computation even if we can't replicate human cognition. In fact, I will argue in the next chapter that we probably don't even want to replicate human cognition in computers. We can do many more useful things with computers that *complement* human capabilities. In chapter 9, I argued that the concept of *semantics* makes the human-computer partnership powerful indeed. Digital technology, together with the stack of hardware and software paradigms that humans have constructed already, is a truly rich medium for modeling, even if the total number of such models is tiny compared with the possibilities. We know how to build faithful physical realizations for almost all of these models. Using Bayesian reasoning, we also know how to model what we don't know about these physical realizations. This gives us a truly expressive toolkit for creativity. I believe we have barely scratched the surface of what we can do with it.

12 Final Thoughts

··· in which I tie it all together, analyze what is holding back technology advancement, and suggest some areas where holding back might not be a bad idea.

12.1 Dualism

The few readers who have gotten this far (thank you!) are probably wondering how I have avoided mentioning Descartes' dualism, the mind-body separation. I've built a whole story about how layers of modeling result in a divorce between software and the physics on which it runs. Software is a model, and I've repeatedly cautioned against confusing the model with the thing being modeled, the map with the territory. A map is a model, and the territory is the physical world being modeled. Isn't my stance the Cartesian dualism all over again, where I've just replaced "mind" with "model"?

A model is physical, in both its physical form as in a paper map and in its mental form, our human conception of a map. Mental states exist in a physical brain. Take away or damage the brain, and the mental states vanish. So even the mental form is physical. Some philosophers call this point of view "physicalism." It holds that all phenomena in the world, including mental states, arise from physical processes. I caution the reader, however, against assuming that this means we can *explain* all physical phenomena. We can't, but even without being able to explain it, everything is physical, and no *fundamental* mind-body dualism exists.

To a scientist, the inability to explain the physical world is profoundly disconcerting. It undermines the positivist agenda. To "explain" a physical phenomenon, according to the positivist philosophy of science, is to construct a (preferably formal or mathematical) model of that phenomenon, a "theory." Following Popper, the theory must be falsifiable by experiment. But Gödel has shown that any formal system capable of self-reference[1] will be either incomplete or inconsistent (chapter 9). Hawking points out that any explanation of physical phenomena takes the form of mental states that live in the very physical world that they explain. Thus, any theory that explains all physical phenomena must be capable of self-reference and

---

1 I speak loosely here when I say "capable of self-reference." The specific requirement is that the system be able to express Gödel's sentence. Gödel showed that any system rich enough to express arithmetic on the natural numbers is rich enough to express Gödel's sentence, and it is hard to imagine trying to use a theory for physics that is not at least this rich.

hence will be either incomplete or inconsistent. Hawking argues that all the models we have of the physical world today are *both* incomplete and inconsistent.

Reinforcing Hawking and Gödel, Wolpert uses a similar self-embedding to conclude that predicting the future is impossible even in a deterministic world (chapter 10). Even basic mathematics becomes suspect. Popper struggled with reconciling the truth of mathematical equations with his falsifiability criterion for science (chapter 1). Consider the equation $1 + 1 = 2$. Can any imaginable experiment falsify this equation? If not, then according to Popper's theory, the equation is not a scientific theory.

Tarski elaborated Gödel's incompleteness to show that no formal system capable of self-reference can define its own notion of truth (chapter 9). If you combine this result with the premise that the physical world can be modeled formally, the essential positivist agenda, then there can be no notion of truth in the world. Positivism collapses into a steaming pile of negativism.

But I am an engineer not a scientist. I work in Simon's "sciences of the artificial" (chapter 1). I am glad some scientists are trying to model the natural world in great detail, but that is *their* agenda not mine. In my world, self-reference is a source of power not a source of weakness. For example, we can use software to build simulation models of semiconductor physics to improve the design of the transistors that ultimately run the software.

I don't need to explain everything in the physical world. In fact, quite the contrary. My agenda is to create things that have never before existed in the physical world. You have to admit that it would be hard to explain things that don't yet exist. I simply don't buy the Platonic extreme which assumes that all those things already exist in some separate reality and are just waiting to be discovered (chapter 1).

The self-scaffolding that arises when there are many layers of separation from the physical reality, as there are in digital technology (chapters 4 and 5), gives us no small measure of freedom to create. As humans, thinking machines, we can *pretend* that we operate in an artificial world of models, separate from physics, use our imagination, and invent clever artifacts and even whole new paradigms. Such new paradigms layer further abstractions on top of the existing ones (chapter 6), furthering the illusion of model-body separation and enabling yet more creativity. Even if those abstractions have limitations, as the notion of computation does (chapter 8), there are far fewer limitations to the semantics that we, as humans, can associate with these abstractions (chapter 9). This bootstrapping of models and the endless possibilities for their meanings are fundamentally powerful and rely on self-reference. Arguably, the very incompleteness of self-referential systems is what *enables* creativity. It ensures that we will never be finished.

Once we recognize that technology is fundamentally a creative enterprise and a partnership between man and machine, then the personalities and idiosyncrasies of the creators of any particular technology become important. We must not treat technologies as dry Platonic facts that have always existed in some other world, waiting to be discovered. Instead, they are cultural, dynamic ideas, subject to fashion, politics, and human foibles. To me, this makes technology much more interesting.

As cultural artifacts, technologies evolve through collective mutation, through design and invention, more than through discovery. Discovery implies an event, the Aha! moment where some preexisting fact becomes known to an individual. The discoverer may be much heralded for a contribution to humanity, but the presumption is that the discoverer as an individual is irrelevant to the fact that has been discovered. Invention and design are not like that. They are less likely to be discrete events, Aha! moments, and they are more likely to come about, like many cultural forces, through the contributions of a large number of people.

To be effective, large numbers of people building collective wisdom must operate within a common mental framework, a "paradigm" to use the word of Thomas Kuhn. Kuhn's paradigms provide the conceptual framework for scientific thought, and Kuhn's theory is that science progresses more through paradigm shifts than through accretion of knowledge. As with science, in technology, common paradigms enable collective development of technologies and interoperability of technologies. Also as with science, paradigm shifts disrupt the equilibria. Accretion of knowledge, normal engineering, coexists with technology revolution, paradigm shift, a process of punctuated equilibrium, to use a term from evolutionary biology.

However, differences in the role of paradigms in science versus engineering do exist. Paradigm shifts are relatively rare in science, for example, Newton's mechanics, Einstein's theory of relativity, and quantum theory. In technology, however, paradigm shifts are relatively frequent. Consider, for example, the richness of styles of programming languages (chapter 5) and their radically different perspectives on computation. Each constitutes a paradigm, and acceptance of that paradigm is necessary to build technologies using those languages.

Of course, a programming language reflects a relatively small paradigm that adheres to a bigger, more durable paradigm. The notion of computation, as developed by Turing, Church, von Neumann, and others during the twentieth century, constitutes a paradigm that persists over time scales more comparable to scientific paradigms. Paradigm shifts coexist with more stable paradigms, a process enabled by the deep layering of paradigms, where one is built on another. There is much less such layering in science, resulting in less frequent and more disruptive paradigm shifts. To be effective, technologists must accept disruption, and to be innovative, technologists must embrace and even seek disruption.

Because of the layering of paradigms, perhaps ironically, the stability of a paradigm at one layer can facilitate change at another layer. For example, because instruction set architectures have changed little in 40 years, they provide a stable platform that decouples the astonishing advances in the underlying semiconductor technology from the creative invention of new programming languages. The relatively stable intermediate layer enables rapid progress in both the layer below and the layer above by insulating these layers from one another.

Scientific paradigms get institutionalized through textbooks, courses, conferences, and a common culture. So do technology paradigms, but in addition, the technology may encode its own paradigms. We particularly see this in software, where, for example, compilers, which translate programs written in a programming language into machine code, can be written in the very languages they compile. This self-scaffolding of paradigms is part of what enables a deep layering because paradigms become precise enough to be built on with confidence.

Digital technologies, layered from the semiconductor physics of transistors up through the sociotechnical phenomenon of Wikipedia and to the "collective, metazoan organism" of a server farm, have been particularly disruptive and powerful for several intertwined reasons. Their discrete and algorithmic nature makes deterministic models (chapter 10) practical and useful. The fact that we can make transistors that switch discretely, on and off, abstracting the messy sloshing of electrons that underlies the physics, enables deterministic, formal, mathematical models with powerful analytical properties and repeatable behaviors. The fact that we can put billions of these transistors into tiny spaces and switch them on and off billions of times per second enables the construction of enormously complex behaviors out of what is ultimately trivially simple, on or off.

The power of digital technologies has created an enthusiasm about them that has spilled over into the world of physics and neuroscience, where legions of serious scholars are convinced that the universe and everything in it, including human cognition, is a Turing computation. But the set of all Turing computations is a tiny set, albeit an infinite one (chapter 8). The sum total of all the things we can do with software is countable, and countable sets are the smallest of all infinite sets that we know about (chapter 9). If we are to accept that nature has for some reason constrained itself to operate only in this smallest of all infinite sets, then we should insist, with great determination, on evidence. The evidence is not there, and we don't even know what sorts of experiments would supply that evidence (chapter 11).

Although digital technologies are severely limited compared with what is likely possible in the physical world (unless we accept digital physics), they are nevertheless exceedingly powerful, particularly when combined with the human notion of semantics (chapter 9). In fact, I claim that there are much more serious obstacles to overcome than the intrinsic incompleteness of these models. I examine these obstacles in the next section.

12.2 Obstacles

The most serious obstacle to overcome is that, despite its amazing capabilities, the human brain is really quite limited. The fact is, we cannot remember as much as computers can. We cannot read as fast. We cannot perform calculations as fast, and we make many more errors. Nevertheless, in partnership with computers and networks, we can do amazing things, such as carry around in our pockets nearly everything humans have ever published.

Digital technology gives humans an astonishingly rich medium for creativity, rich enough that we are nowhere near saturating what can be done with *today's* technology, never mind tomorrow's. Creativity comes at its own rate, however, slowed again by our brain's propensity to be dominated by our unknown knowns (chapter 2). We resist new paradigms, even as the technology makes possible a firehose of new paradigms (chapter 6).

In digital technology, every layer of abstraction except the lowest, semiconductor physics, is a human invention. By the time we get through several layers of abstraction for hardware and then software, we are so many orders of magnitude removed from the physics that physics becomes almost irrelevant (chapter 3). We enter the world of models, human-constructed abstractions. Software becomes more like mathematics and less like natural science, subject only to its own rules. To be sure, we need useful rules, but we need not insist that they not be self-referential. In fact, they *must* be self-referential to be expressive, as Gödel showed.

All of the modeling paradigms that humans have invented so far have limitations. Software can only perform "effective computation," to use Turing's phrase, a tiny subset of the processes that are possible in the physical world (chapter 8). Digital technology exists within only the smallest of the infinite sets identified by Cantor (chapter 9). We can go beyond software and build machines that form a cyberphysical partnership, leveraging the strengths of each (chapter 6). Partnering computers with other physical machines enables us to overcome other human limitations, such as our limited ability to communicate. We write with 10 slow fingers, and we speak using a mere 10 kilohertz of audio bandwidth. Physical machines have no such limitations. Think of the partnerships among artists, computers, digital displays, and the human visual system when you watch a computer-generated movie such as *Avatar*. These partnerships achieve a degree of human-to-human communication never before possible.

We will never be able overcome one killer limitation by ourselves: models are useful to humans only if humans can understand them. Although our brains are truly remarkable machines, they have enormous difficulty with models of even modest complexity and sophistication. There are no doubt models in this book that even the most learned reader will have some difficulty with because the span of specializations that I talk about is so vast. Yet what I talk about in this book barely scratches the surface. The models actually used by specialists in any of the areas I touch on are vastly more sophisticated and cognitively challenging than my oversimplifications might lead you to believe.

Specialization is necessary to enable sophisticated modeling because of our brains' limitations, but it comes at a cost. Kuhn observes that the "paraphernalia of specialization" (specialized journals, professional societies, technical conferences, academic departments, curricula) acquire a prestige of their own and create an inertia against new paradigms (Kuhn, 1962, p. 19). Thus, specialization creates resistance to change.

Because our brains can only fit so much, specialization leads to fragmentation, where insights in one specialty become inaccessible to the others. It can be quite difficult for scientists and engineers to work across specialties. Wulf, writing about Alexander von Humboldt, points out that Humboldt's integrative, cross-disciplinary approach to science went out of fashion (chapter 2), leading to Humboldt being largely forgotten by the scientific community. She observes that "this growing specialization provided a tunnel vision that focused in on ever greater detail, but ignored the global view that would later become Humboldt's hallmark" (Wulf, 2015, p. 22).

With this tunnel vision, specialists know more and more about less and less, until they eventually know everything about nothing. Then they become professors, and the courses they teach become barriers, weeding out unsuspecting undergraduates who simply aren't prepared for the sophistication of the specialty. The professors love their specialty, they want to teach it, and they cannot see that it is esoteric; the arcane and complex analytical methods they have developed are neither easily learned nor easily applied to practical problems. Their discipline fragments into further specialties, and each professor loses the big picture. None is qualified to teach the big picture, and anyway, his or her colleagues would consider any such big picture to be "Mickey Mouse," too easy and unsophisticated to be worthy of their time.

The fragmentation created by specialization is particularly damaging in view of the degree to which technology shapes and pervades our culture. Every person on the planet, regardless of his or her own intellectual predilections, is affected by technology. Yet many intellectuals discount the value of understanding technology. I do not see how a true humanist today can understand society without understanding technology. It seems to me that studying contemporary culture without technology is like studying literature without language. Yet that seems to be what many people do.

My own alma mater, Yale University, a paragon of the arts and sciences, is a case in point. In 1992, a few years after I graduated with a double major in Computer Science and what they called "Engineering and Applied Science," Provost Frank Turner suddenly proposed to eliminate almost all of engineering at Yale. "Applied science" was, to Turner, an intellectual afterthought, not nearly as important as science itself. In his view, applying science was all that engineering did. The closure of engineering didn't happen, but it was dramatic at the time and created quite a firestorm among some of the alumni. Turner was a "historian of the ideas that shaped Western civilization," according to a Yale News memorial published in 2010. Today, I don't see how you could possibly study the ideas that shape Western civilization without studying technology.

Yet technology is not easy to understand. We have probably all had the experience of feeling dumb when we can't figure out something about our computer or home network, and some wizard in India explains to us how to fix the problem. We feel as if we should have known what the wizard knows or should have been able to figure it out. Yet the wizard's knowledge is not about fundamental facts about the world. It is about the language and culture of a specific, idiosyncratic technology. In a pluralistic world, we can't all understand all the languages and cultures that exist. Each specialization has its own parochial gurus.

Specialization gets further amplified by the publish or perish mentality of academia. The most highly regarded publishable results in academic journals are the ones that solve long-standing open problems. In a more specialized field, it is much clearer what the open problems are. The open problems are harder to solve, so you get more respect for solving them. And only a long-standing specialty can have long-standing open problems. In interdisciplinary work and in work in an immature field, by contrast, often the real innovation comes from *formulating* a problem that has not been previously recognized. However, if the newly formulated problem turns out to be easy to solve, then any academic paper describing the solution will likely be rejected under peer review because the solution is "obvious."

Even technology can slow innovation. Software has the rather remarkable property that it tends to encode its own paradigms (chapter 5). When a new software paradigm emerges, it inevitably comes with a suite of languages and tools that quickly accumulate sufficient mass and complexity to become immovable. These make the paradigm durable but sometimes too durable.

On May 25, 2016, the U.S. General Accounting Office released a report stating that more than 70% of the U.S. government's information technology budget is spent on "operations of maintenance" rather than "development, modernization, and enhancement." This amounts to about $65 billion per year, much of which is spent on outdated languages and hardware, some as much as 50 years old. The report cites U.S. Department of Defense use of 8-inch floppy disks and U.S. Treasury Department use of assembly code. Hardware, languages, and tools become an impediment to innovation because change is difficult.

Of course, all disciplines resist change. Kuhn says, "Normal science, for example, often suppresses fundamental novelties because they are necessarily subversive of its basic commitments" (Kuhn, 1962, p. 5). When the "basic commitments" become codified in a suite of million-line computer programs, the inertia can become even harder to overcome.

It is not just the technologists resisting change but also consumers of technology. Humans become used to a means of interaction with machines, and change can become impossible. For example, the brake and accelerator pedals on cars were designed at a time when these controls had direct mechanical and hydraulic couplings with the brakes and throttle. People learned to drive using these controls.

Today, the pedals send commands to a computer, which mediates the commands, possibly changing them before passing them on to the brakes and throttle. For example, the computer may apply different amounts of braking to each wheel to improve stability and prevent skidding. In an electric car, instead of a single throttle, each wheel may have its own electric motor, and the computer may individually control these motors. If humans were to control each wheel with direct mechanical linkages, then the car would have to have eight pedals, four for the brakes and four for the accelerators, and only an octopus could drive.

Car manufacturers could have invented entirely new ways for humans to command the car, for example, using a joystick rather than a steering wheel, but such changes would be resisted by the customers. Humans learned to drive with the old style of controls, and unlearning something is often harder than learning something new. Consequently, car manufacturers today have to work pretty hard to design the pedals so that they feel as if they are connected directly to mechanical and hydraulic actuators. The pedals even push back the way a hydraulic control would, but it is likely that a computer rather than an oil-filled cylinder is determining how much to push back.

A prevalent but misleading view is that human-computer interfaces should be "intuitive." There is nothing intuitive about the pedals in a car. There is nothing intuitive about interacting with a computer. All of the interaction mechanisms we use today are learned. I am reminded of an episode of the TV series, "Star Trek," where the crew of the Enterprise travels back in time to the late 1980s. The engineer, Scotty, needs to use a vintage 1980s computer to solve a problem. So he starts talking to the computer: "Computer. Computer. *Computer!*" The computer, of course, does not respond. One of the vintage 1980s engineers picks up a mouse and hands it to Scotty, saying, "You have to use this." Scotty, looking embarrassed, says, "Oh yes, of course." He picks up the mouse and begins speaking into it as if it were a microphone: "Computer. Computer. *Computer!*" The computer mouse was a brilliant invention, and we have assimilated that paradigm, but there is nothing intuitive about it.

Unfortunately, the way we *teach* technology tends to ignore the inventive nature of technology. I use "we" here to refer to myself and my colleagues, professors at universities and institutes of technology. We teach technology as if it were a collection of facts and truths, Platonic Ideals that exist timeless and independent of humans, waiting to be discovered. A consequence of this style of teaching is that we reinforce the philosophy that values discovery over invention and invention over design (chapter 1).

If what discoverers discover has already existed in the Platonic Good, then the discoverers should not be important as individuals. The discoverers' personalities and predilections cannot possibly have any influence on the nature of the facts and truths they discover. Kuhn observes that this tendency to disconnect ideas from their originators is "ingrained in the ideology of the scientific profession," and he quotes Alfred North Whitehead, saying, "A science that hesitates to forget its founders is lost" (Kuhn, 1962, p. 138). If all worthwhile facts and truths are already out there, waiting to be discovered, then the most valuable contribution an individual can make is simply to bring into light some fact or truth that was previously in darkness.

When talking about technology rather than science, I couldn't disagree more with Whitehead. Almost all "facts and truths" about technology are actually human inventions. As it stands today, technology reflects a chaotic Darwinian evolution of sometimes quirky and idiosyncratic ideas. Understanding the origin of these ideas is essential to being able to think critically about them, and thinking critically about the ideas is essential to technology revolutions.

The fact that we value invention over design can also present an obstacle to creativity. The emphasis on novelty over quality, where "new is better than good," is an academic rathole. No respected academic journal would have published a paper describing the iPhone because every element of the technology already existed in other products. Yet the iPhone was a momentous contribution to humanity, whereas the vast majority of what is published in academic journals is not.

Creative people capable of such contributions may be repelled by technology, perceiving the discipline of engineering as a trade, requiring professional training on well-understood techniques, applying methods known in science, and tweaking and optimizing existing designs (chapter 1). Indeed, much of engineering is "normal engineering" (chapter 6), just as much of science is what Kuhn calls "normal science." Yet most of engineering is quite far from the drudgery that this view implies because it involves creating things that have never before existed. In fact, engineers tend to use their own technology to automate the drudgery, for example, using compilers to translate programs into machine code (chapter 5). I hope I have convinced the reader that engineering is indeed an intellectual and a creative discipline.

Creativity is enabled by the richness of possibilities offered by so many layers of modeling and so many modeling paradigms, but this same richness can overwhelm. It is difficult for humans to comprehend the alternatives, so instead they latch onto those paradigms nearest at hand. By definition, paradigms frame our thinking. Yet in framing our thinking, they also constrain our thinking, limiting our choices. With digital technology, this "freedom from choice," as Sangiovanni-Vincentelli calls it, is essential to designing anything that uses billions of transistors that can switch on and off billions of times per second. Any given individual will have assimilated only a few of the relevant paradigms and will resist straying from these. The resulting fragmentation of the engineering community limits communication between engineers and reinforces sects and their parochial thinking.

It may seem that a reasonable counterforce to this fragmentation is the development of international standards. These can lead to a community coalescing around a common language, principle, or approach. Yet there are costs. One is that the making of international standards is a messy, political, and possibly even corrupt process (chapter 6). Such flawed processes do not usually lead to good technical decisions, particularly when the standard concerns relatively immature technologies.

Standards can also suppress competition among ideas. The U.S. Department of Defense, for example, in an effort to promote standards, for many years forced contractors and researchers to use the Ada programming language, the VHDL hardware description language, and the Microsoft Windows operating system. Ironically, this was occurring while the U.S. Department of Commerce was suing Microsoft for monopolistic practices. In the 1990s, I was involved in a DARPA research project that had the goal of improving the way high-performance signal processing hardware and software was designed, but DARPA required the participants in the project to use VHDL, a particular language for specifying hardware design. However, a language frames the thinking of the designers, so this mandate limited the possible alternatives for innovation. It ruled out one of the key avenues toward innovation in digital technologies, namely, the development of new languages.

Despite many efforts to standardize, digital technology remains a rich and dynamic ecosystem of competing alternative paradigms. In no small part, this is because a given target system or device may have many useful models. For example, a microprocessor chip may be modeled as a three-dimensional geometry of doped silicon, as a computer program specifying software for the chip to run, or anything in between (semiconductor physics, logic gates, instruction set architecture, etc.). These models are all abstractions of the same chip and its function, and they serve different purposes. Which model to use depends on the goal.

Every model is constructed within some modeling paradigm that is typically codified by languages and tools. The languages provide a syntax by which the model is specified (how it is written down or otherwise rendered in physical form) and the semantics (what a given rendition means). The choice of modeling framework has profound consequences. For example, a language for describing three-dimensional shapes is not well suited for modeling the dynamics of a circuit (how the voltages and currents change over time).

Models have many uses, and their intended use should frame the choice of modeling paradigm. They can be used for humans to share ideas asynchronously, like documents. In this case, they need to be simple and use a notation that has been agreed on. Models can be used for specification of a design, in which case simplicity may not be as important as completeness. A computer program that omits a few lines in the name of simplicity will likely not be a useful program. Models can be used for analysis of designs, in which case the formal and mathematical properties of the modeling language can be a dominant concern. Informal languages admit too many possible interpretations to be useful for analysis, even if they are useful for human-to-human communication.

However, humans don't typically choose paradigms. Paradigms are assimilated slowly, often subconsciously, or are drummed in by educators who are likely too specialized to know the alternatives. As a consequence, engineers typically build models using the paradigms they know regardless of whether these are the right choices. This may explain why so many projects fail (chapter 6).

A key challenge is that complex systems, such as the Airbus A350 (chapter 6), have many conflicting modeling requirements. Models need to be combined in ways that their modeling paradigms do not admit easily. A model such as the iron wing prototype of the Airbus A350 (figure 6.4) is able to combine many different aspects of the system in a single model. However, such a model is extremely expensive to construct. If we had better modeling languages and paradigms, then no such physical model would be necessary. A "virtual prototype," which is a reasonably complete model constructed entirely in software, would (likely) be much less expensive. Virtual prototypes are used routinely today for billion-transistor silicon chips, where they work extremely well. Yet for complex cyberphysical systems such as the A350, it is difficult to build enough confidence using virtual prototypes alone.

I remind the reader that models are used differently by engineers and scientists (chapter 2). For an engineer, the things being modeled are expected to imitate the model, whereas for scientists, it is the other way around. Engineers also use models in the scientific way, and scientists use models in the engineering way, but because these uses differ, it is important to understand how a model is being used. A logic gate is almost certainly going to be a poor model of a piece of silicon found in nature, for example, sand on a beach. Yet it is an exceptionally good model for a piece of silicon coming out of an Intel fab.

12.3 Autonomy and Intelligence

Today, some of the most exciting and scary technology developments involve autonomy and intelligence. Both of these terms anthropomorphize computers, a seriously questionable practice (chapter 9). Autonomy refers to the ability that a system has to make decisions without human direction. Intelligence refers to the ability that a system has to exhibit human-like reactions that appear to leverage common knowledge about the world. In both cases, these are not binary properties that are either present or not in systems. Instead, if they are present at all, they are only present in degrees.

Consider self-driving cars, which exist already and will likely be widely available in some form soon. For self-driving cars, full autonomy is not even remotely desirable. Full autonomy would mean that the car accepts no direction at all from humans. You would not even be able to tell it where you want to go. It would simply go where it likes. I don't think a fully autonomous car will sell very well.

How about no autonomy? A car with no autonomy would require you to turn a crank to turn it on. Today, you press a button or turn a key, and a computer tells an electric starter motor to turn the crank for you. This is partial autonomy. It accepts a command from you, an expression of your desires (to turn on the car), and it takes over from there, performing the sequence of actions necessary to turn on the car. Today, that does not even necessarily turn on the engine. A modern braking system is similar, where a computer intervenes to coordinate the braking on the four wheels in a way that no human could do, even if we had four brake pedals.

A few days ago, I saw a flatbed truck go by carrying a nice relatively new car whose front end was completely smashed. This car was not a self-driving model, so almost certainly a human was at fault. My reaction surprised even me. I thought to myself, "That car should be ashamed of itself." Clearly, I was anthropomorphizing the car, but the sentiment was valid. If my thinking had been more rational, I would have said to myself, "The designers of that car should be ashamed of themselves." Today, there really is no excuse for putting cars on the road that will happily crash into the back end of the car in front of them. They should not do this, no matter what their human is telling them to do. We have the technology to solve that problem at a reasonable cost, particularly if we consider the cost of not doing it (e.g., insurance, health, and lives).

A car that refuses to crash is a good example of partial autonomy. The fact is, humans are not very good drivers. They get distracted easily, they get sleepy, they get drunk, and they get old. We have the technology to make much better drivers if we simply give the cars more autonomy.

Yet such a change is not without difficulties. There are ethical questions, for example. How should we program a computer to react when an accident is inevitable? Suppose that the choice is between killing a pedestrian and protecting the passenger or hitting a truck and killing the passenger. It will not do to say that we should program the computer to react as a human would because we don't know how to do that. Whatever software we write for self-driving cars, that software will have hidden in it the outcome of that quandary. Even the software engineers will likely not know what that outcome is.

Consider the goal that Travis Kalanick, CEO of Uber, announced in 2015 to eventually replace their contractor drivers with self-driving cars. There are of course the perennial questions about the impact that such automation has on employment. These questions are difficult, because history has shown that increased automation does not necessarily result in fewer people employed.[2] Consider another possible

---

2 For a scary and pessimistic view of the question of whether technology will finally result in fewer jobs, see Ford (2015).

outcome of such automation. Today, the most effective way to deliver an improvised explosive device (IED) for a terrorist attack is in a car with a suicide bomber. Driverless Uber will eliminate the need for a suicide.

Let's consider an even more difficult question. It is well known that the United States uses unmanned drones as weapons systems, using them, for example, to kill known adversaries. These systems are partially autonomous. At a minimum, they can autonomously navigate to a waypoint circumventing terrain and obstacles. Do they make the decision to launch a lethal weapon autonomously? As far as we know (much of this is hidden behind the veil of classified information), today these decisions are still being made by humans. The humans assess the information obtained from the sensors and actively make the kill decision. Yet we have the technology to give these systems more autonomy.

My colleague at Berkeley Stuart Russell, a noted research leader in artificial intelligence and robotics, has been leading a campaign for an international treaty that would ban such lethal autonomous weapons systems, called LAWS.[3] It is an extremely difficult question how governments and society should react to these technical possibilities, but Russell makes a strong case for such a treaty.

Let's turn our attention to intelligence. First, let me point out that anthropomorphizing computers is not only unjustified by the technology (chapter 9) but is also unreasonable. It is simply not true that we want our machines to exhibit human-like behavior. I really do not want to have to argue with my car about getting to school on time. It's hard enough to have that argument with my daughter.

Consider a Google search of the web. Does Google attempt to give human-like answers? Not really, thankfully. Instead, Google finds answers *written by humans* that are likely to be helpful. Try asking Google, "What is the meaning of life?" When I did this just now (May 29, 2016), the first hit on the list of possibly helpful pages is a link to a wonderful Wikipedia page on the subject. That page even includes a discussion of the answer "42" to this question (see footnote on page 84). Google is brokering for me the collective intelligence of humans. In my opinion, this does not in any way replace human intelligence. Quite the contrary, it augments human intelligence. I'm quite sure it makes me smarter because I have a really poor memory, and it improves the ability of humans to communicate with one another by democratizing publication. Everyone has a voice.

If we accept digital physics and the universality of software, which I don't (chapter 8), then in principle we should be able to make computers that will make decisions at least as well as humans, for example, the kill decisions of LAWS. Digital physics implies that computers can be endowed with the same sense of identity, self-awareness, accountability, and feelings as a human because all of these things

---

3 See http://www.cs.berkeley.edu/~russell/research/LAWS.html

must be digitally replicable. But why would we want to delegate such decisions to computers that emulate humans? We have seven billion human brains on the planet as it is. Isn't that enough? We should instead be focusing on the ways in which technology complements human capabilities. Humans can keep seven numbers in short-term memory. The smartphone in your pocket can keep billions. Some decisions can be made *better* by computers, not by replicating human processes, but by exploiting their ability to evaluate millions or even billions of alternatives.

If we actually succeed in making computers behave like humans, I'm quite certain we will come to regret it. Humans, after all, don't really behave very well. What kinds of new wars will we find ourselves in? What if the computers decide to take humans out of the loop, as they do in *The Matrix*, the science fiction film trilogy?

Autonomy and intelligence are about getting computers to do more for us. If we can orchestrate things so that what the computers do for us, on net, provides real benefits, then we come out ahead. However, it is unreasonable to expect there to be no costs. For this reason, technology development should not be left exclusively to the nerds, like me. I believe it is imperative for nonnerds to understand technology better so they can help guide its evolution in society, and it is imperative for the nerds to understand the cultural context of their work. These are the main reasons I wrote this book.

Bibliography

Abbott, B. P., R. Abbott, T. D. Abbott, M. R. Abernathy, F. Acernese, K. Ackley, C. Adams, T. Adams, P. Addesso, R. X. Adhikari, V. B. Adya, C. Affeldt, M. Agathos, K. Agatsuma, N. Aggarwal, O. D. Aguiar, L. Aiello, A. Ain, P. Ajith, B. Allen, A. Allocca, P. A. Altin, S. B. Anderson, W. G. Anderson, K. Arai, M. A. Arain, M. C. Araya, C. C. Arceneaux, J. S. Areeda, N. Arnaud, K. G. Arun, S. Ascenzi, G. Ashton, M. Ast, S. M. Aston, P. Astone, P. Aufmuth, C. Aulbert, S. Babak, P. Bacon, M. K. M. Bader, P. T. Baker, F. Baldaccini, G. Ballardin, S. W. Ballmer, J. C. Barayoga, S. E. Barclay, B. C. Barish, D. Barker, F. Barone, B. Barr, L. Barsotti, M. Barsuglia, D. Barta, J. Bartlett, M. A. Barton, I. Bartos, R. Bassiri, A. Basti, J. C. Batch, C. Baune, V. Bavigadda, M. Bazzan, B. Behnke, M. Bejger, C. Belczynski, A. S. Bell, C. J. Bell, B. K. Berger, J. Bergman, G. Bergmann, C. P. L. Berry, D. Bersanetti, A. Bertolini, J. Betzwieser, S. Bhagwat, R. Bhandare, I. A. Bilenko, G. Billingsley, J. Birch, R. Birney, O. Birnholtz, S. Biscans, A. Bisht, M. Bitossi, C. Biwer, M. A. Bizouard, J. K. Blackburn, C. D. Blair, D. G. Blair, R. M. Blair, S. Bloemen, O. Bock, T. P. Bodiya, M. Boer, G. Bogaert, C. Bogan, A. Bohe, P. Bojtos, C. Bond, F. Bondu, R. Bonnand, B. A. Boom, R. Bork, V. Boschi, S. Bose, Y. Bouffanais, A. Bozzi, C. Bradaschia, P. R. Brady, V. B. Braginsky, M. Branchesi, J. E. Brau, T. Briant, A. Brillet, M. Brinkmann, V. Brisson, P. Brockill, A. F. Brooks, D. A. Brown, D. D. Brown, N. M. Brown, C. C. Buchanan, A. Buikema, T. Bulik, H. J. Bulten, A. Buonanno, D. Buskulic, C. Buy, R. L. Byer, M. Cabero, L. Cadonati, G. Cagnoli, C. Cahillane, J. C. Bustillo, T. Callister, E. Calloni, J. B. Camp, K. C. Cannon, J. Cao, C. D. Capano, E. Capocasa, F. Carbognani, S. Caride, J. C. Diaz, C. Casentini, S. Caudill, M. Cavaglià, F. Cavalier, R. Cavalieri, G. Cella, C. B. Cepeda, L. C. Baiardi, G. Cerretani, E. Cesarini, R. Chakraborty, T. Chalermsongsak, S. J. Chamberlin, M. Chan, S. Chao, P. Charlton, E. Chassande-Mottin, H. Y. Chen, Y. Chen, C. Cheng, A. Chincarini, A. Chiummo, H. S. Cho, M. Cho, J. H. Chow, N. Christensen, Q. Chu, S. Chua, S. Chung, G. Ciani, F. Clara, J. A. Clark, F. Cleva, E. Coccia, P.-F. Cohadon, A. Colla, C. G. Collette, L. Cominsky, M. Constancio, A. Conte, L. Conti, D. Cook, T. R. Corbitt, N. Cornish, A. Corsi, S. Cortese, C. A. Costa, M. W. Coughlin, S. B. Coughlin, J.-P. Coulon, S. T. Countryman, P. Couvares, E. E. Cowan, D. M. Coward, M. J. Cowart, D. C. Coyne, R. Coyne, K. Craig, J. D. E. Creighton, T. D. Creighton, J. Cripe, S. G. Crowder, A. M. Cruise, A. Cumming, L. Cunningham, E. Cuoco, T. D. Canton, S. L. Danilishin, S. D'Antonio, K. Danzmann, N. S. Darman, C. F. Da Silva Costa, V. Dattilo, I. Dave, H. P. Daveloza, M. Davier, G. S. Davies, E. J. Daw, R. Day, S. De, D. DeBra, G. Debreczeni, J. Degallaix, M. De Laurentis, S. Deléglise, W. Del Pozzo, T. Denker, T. Dent, H. Dereli, V. Dergachev, R. T. DeRosa, R. De Rosa, R. DeSalvo, S. Dhurandhar, M. C. Díaz, L. Di Fiore, M. Di Giovanni, A. Di Lieto, S. Di Pace, I. Di Palma, A. Di Virgilio, G. Dojcinoski, V. Dolique, F. Donovan, K. L. Dooley, S. Doravari, R. Douglas, T. P. Downes, M. Drago, R. W. P. Drever, J. C. Driggers, Z. Du, M. Ducrot, S. E. Dwyer, T. B. Edo, M. C. Edwards, A. Effler, H.-B. Eggenstein, P. Ehrens, J. Eichholz, S. S. Eikenberry, W. Engels, R. C. Essick, T. Etzel, M. Evans, T. M. Evans, R. Everett, M. Factourovich, V. Fafone, H. Fair, S. Fairhurst, X. Fan, Q. Fang, S. Farinon, B. Farr, W. M. Farr, M. Favata, M. Fays, H. Fehrmann, M. M. Fejer, D. Feldbaum, I. Ferrante, E. C. Ferreira, F. Ferrini, F. Fidecaro, L. S. Finn, I. Fiori, D. Fiorucci, R. P. Fisher, R. Flaminio, M. Fletcher, H. Fong, J.-D. Fournier, S. Franco, S. Frasca, F. Frasconi, M. Frede, Z. Frei, A. Freise, R. Frey, V. Frey, T. T. Fricke, P. Fritschel, V. V. Frolov, P. Fulda, M. Fyffe, H. A. G. Gabbard, J. R. Gair, L. Gammaitoni, S. G. Gaonkar, F. Garufi, A. Gatto, G. Gaur, N. Gehrels, G. Gemme, B. Gendre, E. Genin, A. Gennai, J. George, L. Gergely, V. Germain, A. Ghosh, A. Ghosh, S. Ghosh, J. A. Giaime, K. D. Giardina, A. Giazotto, K. Gill, A. Glaefke, J. R. Gleason, E. Goetz, R. Goetz, L. Gondan, G. González, J. M. G. Castro, A. Gopakumar, N. A. Gordon, M. L. Gorodetsky, S. E. Gossan, M. Gosselin, R. Gouaty, C. Graef, P. B. Graff, M. Granata, A. Grant, S. Gras, C. Gray, G. Greco, A. C. Green, R. J. S. Greenhalgh, P. Groot, H. Grote, S. Grunewald, G. M. Guidi, X. Guo, A. Gupta, M. K. Gupta, K. E. Gushwa, E. K. Gustafson, R. Gustafson, J. J. Hacker, B. R. Hall, E. D. Hall, G. Hammond, M. Haney, M. M. Hanke, J. Hanks, C. Hanna, M. D. Hannam, J. Hanson, T. Hardwick, J. Harms, G. M. Harry, I. W. Harry, M. J. Hart, M. T. Hartman, C.-J. Haster, K. Haughian, J. Healy, A. Heidmann, M. C. Heintze, G. Heinzel, H. Heitmann, P. Hello, G. Hemming, M. Hendry, I. S. Heng, J. Hennig, A. W. Heptonstall, M. Heurs, S. Hild, D. Hoak, K. A. Hodge, D. Hofman, S. E. Hollitt, K. Holt, D. E. Holz, P. Hopkins, D. J. Hosken, J. Hough, E. A. Houston, E. J. Howell, Y. M. Hu, S. Huang, E. A. Huerta, D. Huet, B. Hughey, S. Husa, S. H. Huttner, T. Huynh-Dinh, A. Idrisy, N. Indik, D. R. Ingram, R. Inta, H. N.

Isa, J.-M. Isac, M. Isi, G. Islas, T. Isogai, B. R. Iyer, K. Izumi, M. B. Jacobson, T. Jacqmin, H. Jang, K. Jani, P. Jaranowski, S. Jawahar, F. Jiménez-Forteza, W. W. Johnson, N. K. Johnson-McDaniel, D. I. Jones, R. Jones, R. J. G. Jonker, L. Ju, K. Haris, C. V. Kalaghatgi, V. Kalogera, S. Kandhasamy, G. Kang, J. B. Kanner, S. Karki, M. Kasprzack, E. Katsavounidis, W. Katzman, S. Kaufer, T. Kaur, K. Kawabe, F. Kawazoe, F. Kéfélian, M. S. Kehl, D. Keitel, D. B. Kelley, W. Kells, R. Kennedy, D. G. Keppel, J. S. Key, A. Khalaidovski, F. Y. Khalili, I. Khan, S. Khan, Z. Khan, E. A. Khazanov, N. Kijbunchoo, C. Kim, J. Kim, K. Kim, N.-G. Kim, N. Kim, Y.-M. Kim, E. J. King, P. J. King, D. L. Kinzel, J. S. Kissel, L. Kleybolte, S. Klimenko, S. M. Koehlenbeck, K. Kokeyama, S. Koley, V. Kondrashov, A. Kontos, S. Koranda, M. Korobko, W. Z. Korth, I. Kowalska, D. B. Kozak, V. Kringel, B. Krishnan, A. Królak, C. Krueger, G. Kuehn, P. Kumar, R. Kumar, L. Kuo, A. Kutynia, P. Kwee, B. D. Lackey, M. Landry, J. Lange, B. Lantz, P. D. Lasky, A. Lazzarini, C. Lazzaro, P. Leaci, S. Leavey, E. O. Lebigot, C. H. Lee, H. K. Lee, H. M. Lee, K. Lee, A. Lenon, M. Leonardi, J. R. Leong, N. Leroy, N. Letendre, Y. Levin, B. M. Levine, T. G. F. Li, A. Libson, T. B. Littenberg, N. A. Lockerbie, J. Logue, A. L. Lombardi, L. T. London, J. E. Lord, M. Lorenzini, V. Loriette, M. Lormand, G. Losurdo, J. D. Lough, C. O. Lousto, G. Lovelace, H. Lück, A. P. Lundgren, J. Luo, R. Lynch, Y. Ma, T. MacDonald, B. Machenschalk, M. MacInnis, D. M. Macleod, F. Magaña Sandoval, R. M. Magee, M. Mageswaran, E. Majorana, I. Maksimovic, V. Malvezzi, N. Man, I. Mandel, V. Mandic, V. Mangano, G. L. Mansell, M. Manske, M. Mantovani, F. Marchesoni, F. Marion, S. Márka, Z. Márka, A. S. Markosyan, E. Maros, F. Martelli, L. Martellini, I. W. Martin, R. M. Martin, D. V. Martynov, J. N. Marx, K. Mason, A. Masserot, T. J. Massinger, M. Masso-Reid, F. Matichard, L. Matone, N. Mavalvala, N. Mazumder, G. Mazzolo, R. McCarthy, D. E. McClelland, S. McCormick, S. C. McGuire, G. McIntyre, J. McIver, D. J. McManus, S. T. McWilliams, D. Meacher, G. D. Meadors, J. Meidam, A. Melatos, G. Mendell, D. Mendoza-Gandara, R. A. Mercer, E. Merilh, M. Merzougui, S. Meshkov, C. Messenger, C. Messick, P. M. Meyers, F. Mezzani, H. Miao, C. Michel, H. Middleton, E. E. Mikhailov, L. Milano, J. Miller, M. Millhouse, Y. Minenkov, J. Ming, S. Mirshekari, C. Mishra, S. Mitra, V. P. Mitrofanov, G. Mitselmakher, R. Mittleman, A. Moggi, M. Mohan, S. R. P. Mohapatra, M. Montani, B. C. Moore, C. J. Moore, D. Moraru, G. Moreno, S. R. Morriss, K. Mossavi, B. Mours, C. M. Mow-Lowry, C. L. Mueller, G. Mueller, A. W. Muir, A. Mukherjee, D. Mukherjee, S. Mukherjee, N. Mukund, A. Mullavey, J. Munch, D. J. Murphy, P. G. Murray, A. Mytidis, I. Nardecchia, L. Naticchioni, R. K. Nayak, V. Necula, K. Nedkova, G. Nelemans, M. Neri, A. Neunzert, G. Newton, T. T. Nguyen, A. B. Nielsen, S. Nissanke, A. Nitz, F. Nocera, D. Nolting, M. E. N. Normandin, L. K. Nuttall, J. Oberling, E. Ochsner, J. O'Dell, E. Oelker, G. H. Ogin, J. J. Oh, S. H. Oh, F. Ohme, M. Oliver, P. Oppermann, R. J. Oram, B. O'Reilly, R. O'Shaughnessy, C. D. Ott, D. J. Ottaway, R. S. Ottens, H. Overmier, B. J. Owen, A. Pai, S. A. Pai, J. R. Palamos, O. Palashov, C. Palomba, A. Pal-Singh, H. Pan, Y. Pan, C. Pankow, F. Pannarale, B. C. Pant, F. Paoletti, A. Paoli, M. A. Papa, H. R. Paris, W. Parker, D. Pascucci, A. Pasqualetti, R. Passaquieti, D. Passuello, B. Patricelli, Z. Patrick, B. L. Pearlstone, M. Pedraza, R. Pedurand, L. Pekowsky, A. Pele, S. Penn, A. Perreca, H. P. Pfeiffer, M. Phelps, O. Piccinni, M. Pichot, M. Pickenpack, F. Piergiovanni, V. Pierro, G. Pillant, L. Pinard, I. M. Pinto, M. Pitkin, J. H. Poeld, R. Poggiani, P. Popolizio, A. Post, J. Powell, J. Prasad, V. Predoi, S. S. Premachandra, T. Prestegard, L. R. Price, M. Prijatelj, M. Principe, S. Privitera, R. Prix, G. A. Prodi, L. Prokhorov, O. Puncken, M. Punturo, P. Puppo, M. Pürrer, H. Qi, J. Qin, V. Quetschke, E. A. Quintero, R. Quitzow-James, F. J. Raab, D. S. Rabeling, H. Radkins, P. Raffai, S. Raja, M. Rakhmanov, C. R. Ramet, P. Rapagnani, V. Raymond, M. Razzano, V. Re, J. Read, C. M. Reed, T. Regimbau, L. Rei, S. Reid, D. H. Reitze, H. Rew, S. D. Reyes, F. Ricci, K. Riles, N. A. Robertson, R. Robie, F. Robinet, A. Rocchi, L. Rolland, J. G. Rollins, V. J. Roma, J. D. Romano, R. Romano, G. Romanov, J. H. Romie, D. Rosińska, S. Rowan, A. Rüdiger, P. Ruggi, K. Ryan, S. Sachdev, T. Sadecki, L. Sadeghian, L. Salconi, M. Saleem, F. Salemi, A. Samajdar, L. Sammut, L. M. Sampson, E. J. Sanchez, V. Sandberg, B. Sandeen, G. H. Sanders, J. R. Sanders, B. Sassolas, B. S. Sathyaprakash, P. R. Saulson, O. Sauter, R. L. Savage, A. Sawadsky, P. Schale, R. Schilling, R. Schmidt, P. Schmidt, R. Schnabel, R. M. S. Schofield, A. Schönbeck, E. Schreiber, D. Schuette, B. F. Schutz, J. Scott, S. M. Scott, D. Sellers, A. S. Sengupta, D. Sentenac, V. Sequino, A. Sergeev, G. Serna, Y. Setyawati, A. Sevigny, D. A. Shaddock, T. Shaffer, S. Shah, M. S. Shahriar, M. Shaltev, Z. Shao, B. Shapiro, P. Shawhan, A. Sheperd, D. H. Shoemaker, D. M. Shoemaker, K. Siellez, X. Siemens, D. Sigg, A. D. Silva, D. Simakov, A. Singer, L. P. Singer, A. Singh, R. Singh, A. Singhal, A. M. Sintes, B. J. J. Slagmolen, J. R. Smith, M. R. Smith, N. D. Smith, R. J. E. Smith, E. J. Son, B. Sorazu, F. Sorrentino, T. Souradeep, A. K. Srivastava, A. Staley, M. Steinke, J. Steinlechner, S. Steinlechner, D. Steinmeyer, B. C. Stephens, S. P. Stevenson, R. Stone, K. A. Strain, N. Straniero, G. Stratta, N. A. Strauss, S. Strigin, R. Sturani, A. L. Stuver, T. Z. Summerscales, L. Sun, P. J. Sutton, B. L. Swinkels, M. J. Szczepańczyk, M. Tacca, D. Talukder, D. B. Tanner, M. Tápai, S. P. Tarabrin, A. Taracchini, R. Taylor, T. Theeg, M. P. Thirugnanasambandam, E. G. Thomas,

M. Thomas, P. Thomas, K. A. Thorne, K. S. Thorne, E. Thrane, S. Tiwari, V. Tiwari, K. V. Tokmakov, C. Tomlinson, M. Tonelli, C. V. Torres, C. I. Torrie, D. Töyrä, F. Travasso, G. Traylor, D. Trifirò, M. C. Tringali, L. Trozzo, M. Tse, M. Turconi, D. Tuyenbayev, D. Ugolini, C. S. Unnikrishnan, A. L. Urban, S. A. Usman, H. Vahlbruch, G. Vajente, G. Valdes, M. Vallisneri, N. van Bakel, M. van Beuzekom, J. F. J. van den Brand, C. Van Den Broeck, D. C. Vander-Hyde, L. van der Schaaf, J. V. van Heijningen, A. A. van Veggel, M. Vardaro, S. Vass, M. Vasúth, R. Vaulin, A. Vecchio, G. Vedovato, J. Veitch, P. J. Veitch, K. Venkateswara, D. Verkindt, F. Vetrano, A. Viceré, S. Vinciguerra, D. J. Vine, J.-Y. Vinet, S. Vitale, T. Vo, H. Vocca, C. Vorvick, D. Voss, W. D. Vousden, S. P. Vyatchanin, A. R. Wade, L. E. Wade, M. Wade, S. J. Waldman, M. Walker, L. Wallace, S. Walsh, G. Wang, H. Wang, M. Wang, X. Wang, Y. Wang, H. Ward, R. L. Ward, J. Warner, M. Was, B. Weaver, L.-W. Wei, M. Weinert, A. J. Weinstein, R. Weiss, T. Welborn, L. Wen, P. Weßels, T. Westphal, K. Wette, J. T. Whelan, S. E. Whitcomb, D. J. White, B. F. Whiting, K. Wiesner, C. Wilkinson, P. A. Willems, L. Williams, R. D. Williams, A. R. Williamson, J. L. Willis, B. Willke, M. H. Wimmer, L. Winkelmann, W. Winkler, C. C. Wipf, A. G. Wiseman, H. Wittel, G. Woan, J. Worden, J. L. Wright, G. Wu, J. Yablon, I. Yakushin, W. Yam, H. Yamamoto, C. C. Yancey, M. J. Yap, H. Yu, M. Yvert, A. Zadrożny, L. Zangrando, M. Zanolin, J.-P. Zendri, M. Zevin, F. Zhang, L. Zhang, M. Zhang, Y. Zhang, C. Zhao, M. Zhou, Z. Zhou, X. J. Zhu, M. E. Zucker, S. E. Zuraw, and J. Zweizig, 2016: Observation of gravitational waves from a binary black hole merger. *Phys. Rev. Lett.*, **116**. doi:10.1103/PhysRevLett.116.061102.

Abelson, H. and G. J. Sussman, 1996: *Structure and Interpretation of Computer Programs*. MIT Press, Cambridge, MA, 2nd ed.

Akana, J., D. J. Coster, D. D. Iuliis, E. Hankey, R. P. Howarth, J. P. Ive, S. Jobs, D. R. Kerr, S. Nishibori, M. D. Rohrbach, P. Russell-Clarke, C. J. Stringer, E. A. Whang, and R. Zorkendorfer, 2012: Portable display device. US Patent D670,286. Available from: https://www.google.com/patents/USD670286.

Arvind, L. Bic, and T. Ungerer, 1991: Evolution of data-flow computers. In Gaudiot, J.-L. and L. Bic, eds., *Advanced Topics in Data-Flow Computing*, Prentice-Hall, Upper Saddle River, NJ.

Barry, J. R., E. A. Lee, and D. G. Messerschmitt, 2004: *Digital Communication*. Springer Science + Business Media, LLC, New York, 3rd ed.

Bekenstein, J. D., 1973: Black holes and entropy. *Physical Review D*, **7**(8), 2333–2346. doi:10.1103/PhysRevD.7.2333.

Bickle, J., 2016: Multiple realizability. *The Stanford Encyclopedia of Philosophy*, Spring 2016 Edition, Edward N. Zalta (ed).

Binder, P.-M., 2008: Theories of almost everything. *Nature*, **455**(7215), 884–885. doi:10.1038/455884a.

Blum, L., M. Shub, and S. Smale, 1989: On a theory of computation and complexity over the real numbers: NP-completeness, recursive functions and universal machines. *Bulletin (New Series) of the American Mathematical Society*, **21**(1).

Box, G. E. P. and N. R. Draper, 1987: *Empirical Model-Building and Response Surfaces*. Wiley Series in Probability and Statistics, Wiley, Hoboken, NJ.

Britcher, R. N., 1999: *The Limits of Software: People, Projects, and Perspectives*. Addison-Wesley, Reading, MA.

Brooks, F. P., 1975: *The Mythical Man-Month: Essays on Software Engineering*. Addison-Wesley, Reading, MA.

—, 1987: No silver bullet — essence and accidents of software engineering. *Computer*, **20**(4), 10–19. doi:10.1109/MC.1987.1663532.

Bush, V., 1945: As we may think. *ACM SIGPC Notes*, **1**(4), 36–44, reprinted from The Atlantic Monthly, 1945. doi:10.1145/1113634.1113638.

Chaitin, G., 2005: *Meta Math! — The Quest for Omega*. Vintage Books, New York.

Chandra, V., 2014: *Geek Sublime — The Beauty of Code, the Code of Beauty*. Graywolf Press, Minneapolis, MN.

Choi, Y.-K., N. Lindert, P. Xuan, S. Tang, D. Ha, E. Anderson, T.-J. King, J. Bokor, and C. Hu, 2001: Sub-20nm CMOS FinFET technologies. *IEEE International Electron Devices Meeting (IEDM) Technical Digest*, 421–424. Available from: http://koasas.kaist.ac.kr/handle/10203/573.

Cone, E., 2002: The ugly history of tool development at the FAA. Baseline (Blog). Available from: http://www.baselinemag.com/c/a/Projects-Processes/The-Ugly-History-of-Tool-Development-at-the-FAA.

Connor, S., 2009: Michel Serres: The hard and the soft. Report, transcript of a talk given at the Centre for Modern Studies, University of York. Available from: http://stevenconnor.com/hardsoft/hardsoft.pdf.

Daugman, J. G., 2001: Brain metaphor and brain theory. In Bechtel, W. P., P. Mandik, J. Mundale, and R. S. Stufflebeam, eds., *Philosophy and the Neurosciences: A Reader*, Blackwell, Oxford.

Davis, M., 2006: Why there is no such discipline as hypercomputation. *Mathematics and Computation*, **178**, 2–7. doi:10.1016/j.amc.2005.09.066.

Deutsch, D., 2011: *The Beginning of Infinity — Explanations that Transform the World*. Viking, New York.

—, 2012: Creative blocks — the very laws of physics imply that artificial intelligence must be possible. What's holding us up? *Aeon (online magazine)*. Available from: https://aeon.co/essays/how-close-are-we-to-creating-artificial-intelligence.

Dijkstra, E. W., 1972: The humble programmer. *Communications of the ACM*, **15(2)**, 859–866.

Dyson, G., 2012: *Turing's Cathedral — The Origins of the Digial Universe*. Pantheon Books, New York.

Earman, J., 1986: *A Primer on Determinism*, vol. 32 of *The University of Ontario Series in Philosophy of Science*. D. Reidel Publishing Company, Dordrecht, Holland.

Endy, D., 2005: Foundations for engineering biology. *Nature*, **438(7067)**, 449–453.

Field, E. H. and K. R. Milner, 2008: Forecasting California's earthquakes — what can we expect in the next 30 years? *U.S. Geological Survey*, USGS Fact Sheet 2008-3027, from the 2007 Working Group on California Earthquake Probabilities. Available from: http://pubs.usgs.gov/fs/2008/3027/.

Fisher, J., N. Piterman, and M. Y. Vardi, 2011: The only way is up. In Butler, M. and W. Schulte, eds., *Formal Methods (FM)*, Springer-Verlag, vol. LNCS 6664, pp. 3–11.

Ford, M., 2015: *Rise of the Robots — Technology and the Threat of a Jobless Future*. Basic Books, New York.

Franzén, T., 2005: *Gödel's Theorem: An Incomplete Guide to Its Use and Abuse*. A K Peters Ltd., Wellesley, MA.

Freiberger, M., 2014: The limits of information. *+plus Magazine*, Retrieved May 24, 2016. Available from: https://plus.maths.org/content/bekenstein.

Gamma, E., R. Helm, R. Johnson, and J. Vlissides, 1994: *Design Patterns: Elements of Reusable Object-Oriented Software*. Addison Wesley, Reading, MA.

Golomb, S. W., 1967: *Shift Register Sequences*. Aegean Park Press, Laguna Hills, CA.

—, 1971: Mathematical models: Uses and limitations. *IEEE Transactions on Reliability*, **R-20(3)**, 130–131. doi:10.1109/TR.1971.5216113.

Harris, S., 2012: *Free Will*. Free Press, New York.

Hartley, R. V. L., 1928: Transmission of information. *Bell System Technical Journal*, **7(3)**, 535–563. doi:10.1002/j.1538-7305.1928.tb01236.x.

Hawking, S., 2002: Gödel and the end of the universe. *Stephen Hawking Public Lectures*. Available from: http://www.hawking.org.uk/godel-and-the-end-of-physics.html.

Heffernan, V., 2016: *Magic and Loss — The Internet as Art*. Simon & Schuster, New York.

Heitin, L., 2015: When did science education become STEM? *Education Week Blogs*. Available from: http://blogs.edweek.org/edweek/curriculum/2015/04/when_did_science_education_become_STEM.html.

Hennessy, J. L. and D. A. Patterson, 1990: *Computer Architecture: A Quantitative Approach*. Morgan Kaufmann, Burlington, MA.

Hoefer, C., 2016: Causal determinism. *The Stanford Encyclopedia of Philosophy*, Spring 2016 Edition. Available from: http://plato.stanford.edu/archives/spr2016/entries/determinism-causal/.

Hofstadter, D., 1979: *Gödel, Escher and Bach: An Eternal Golden Braid*. Basic Books, New York.

Horgan, J., 1992: Claude E. Shannon [profile]. *IEEE Spectrum*, **29(4)**, 72–75. doi:10.1109/MSPEC.1992.672257.

—, 2016: Is the gravitational-wave claim true? And was it worth the cost? *Scientific American, Cross-Check Column*. Available from: http://blogs.scientificamerican.com/cross-check/is-the-gravitational-wave-claim-true-and-was-it-worth-the-cost/.

IBM, 1968: System/360 model 25. *IBM Archives*, retrieved March 20, 2016. Available from: https://www-03.ibm.com/ibm/history/exhibits/mainframe/mainframe_PP2025.html.

Klaw, S., 1968: *The New Brahmins — Scientific Life in America*. William Morrow & Company, New York.

Kline, M., 1980: *Mathematics — The Loss of Certainty*. Oxford University Press, Oxford, England.

Knuth, D. E., 1984: Literate programming. *The Computer Journal*, **27(2)**, 97–111. doi:10.1093/comjnl/27.2.97.

Kuhn, T. S., 1962: *The Structure of Scientific Revolutions.*. University of Chicago Press, Chicago, IL.

Lakatos, I., 1970: Falsification and the methodology of scientific research programs. In Lakatos, I. and A. Musgrave, eds., *Criticism and the Growth of Knowledge*, Cambridge University Press, Proceedings of the International Colloquium in the Philosophy of Science, London, 1965.

Laplace, P.-S., 1901: *A Philosophical Essay on Probabilities*. John Wiley and Sons, Hoboken, NJ, translated from the sixth French edition by F. W. Truscott and F. L. Emory.

Lee, E. A., 2016: Fundamental limits of cyber-physical systems modeling. *ACM Transactions on Cyber-Physical Systems*, 1(1), 26. doi:10.1145/2912149.

Lee, E. A. and P. Varaiya, 2011: *Structure and Interpretation of Signals and Systems*. LeeVaraiya.org, 2nd ed. Available from: http://LeeVaraiya.org.

Lerdorf, R., 2003: PHP. Interview in IT Conversations, Behind the Mic Series. Available from: https://web.archive.org/web/20130728125152/http://itc.conversationsnetwork.org/shows/detail58.html.

Leuf, B. and W. Cunningham, 2001: *The Wiki Way — Quick Collaboration on the Web*. Addison Wesley, Reading, MA.

Lilienfeld, J. E., 1930: Method and apparatus for controlling electric currents. U.S. Patent 1,745,175. Available from: https://www.google.com/patents/US1745175.

Lloyd, S., 2006: *Programming the Universe—A Quantum Computer Scientist Takes On the Cosmos*. Alfred A. Knopf, New York.

Lorenz, E. N., 1963: Deterministic nonperiodic flow. *Journal of the Atmospheric Sciences*, **20**, 130–141.

Macilwain, C., 2010: Scientists vs engineers: this time it's financial. *Nature*, **467(885)**. doi:10.1038/467885a.

Moskowitz, E., 2016: The chirp that proved Einstein right. *Boston Globe*, May 15, 2016.

NASA, 2016: Spinoff. Report, NASA Technology Transfer Program. Available from: http://spinoff.nasa.gov.

Nichols, S., 2006: Why was Humboldt forgotten in the United States? *Geographical Review*, **96(3)**, 399–415, *Humboldt in the Americas*, published by the American Geographical Society. Available from: http://www.jstor.org/stable/30034515.

Overbye, D., 2016: Gravitational waves detected, confirming Einstein's theory. *New York Times*. Available from: http://nyti.ms/1SKjTJ5.

Pappas, S., 2016: How big is the internet, really? *Live Science*. Available from: http://www.livescience.com/54094-how-big-is-the-internet.html.

Patterson, D. A. and J. L. Hennessy, 1996: *Computer Architecture: A Quantitative Approach*. Morgan Kaufmann, Burlington, MA, 2nd ed.

Penrose, R., 1989: *The Emperor's New Mind—Concerning Computers, Minds and the Laws of Physics*. Oxford University Press, Oxford.

Pernin, C. G., E. Axelband, J. A. Drezner, B. B. Dille, J. Gordon IV, B. J. Held, K. S. McMahon, W. L. Perry, C. Rizzi, A. R. Shah, P. A. Wilson, and J. M. Sollinge, 2012: Lessons from the Army's Future Combat Systems program. Report, RAND Corporation. Available from: http://www.rand.org/content/dam/rand/pubs/monographs/2012/RAND_MG1206.pdf.

Piccinini, G., 2007: Computational modelling vs. computational explanation: Is everything a Turing machine, and does it matter to the philosophy of mind? *Australasian Journal of Philosophy*, **85(1)**, 93–115. doi:10.1080/00048400601176494.

Popper, K., 1959: *The Logic of Scientific Discovery*. Hutchinson & Co., Taylor & Francis edition, 2005, London and New York.

Raatikainen, P., 2015: Gödel's incompleteness theorems. *The Stanford Encyclopedia of Philosophy*, Spring 2015 edition, Edward N. Zalta, Editor. Available from: http://plato.stanford.edu/archives/spr2015/entries/goedel-incompleteness/.

Read, L. E., 1958: I pencil: My family tree as told to Leonard E. Reed. *The Freeman*, republished in 1999 by Irvington-on-Hudson, New York: Foundation for Economic Education, Inc. Available from: http://oll.libertyfund.org/titles/112.

Redford, J., 2012: The physics of God and the quantum gravity theory of everything. *Social Science Research Network (SSRN)*. doi:10.2139/ssrn.1974708.

Rheingold, H., 2000: *Tools for Thought—The History and Future of Mind-Expanding Technology*. MIT Press, Cambridge, MA, first published in 1985 by Simon & Schuster/Prentice Hall.

Rozmanith, A. I. and N. Berinson, 1993: Remote query communication system. U.S. Patent 5,253,341. Available from: https://www.google.com/patents/US5253341.

Rumsfeld, D. H., 2002: DoD news briefing: Secretary Rumsfeld and Gen. Myers. *News Transcript*, U.S. Department of Defense. Available from: http://archive.defense.gov/Transcripts/Transcript.aspx?TranscriptID=2636.

Saint, N., 2010: Google press conference at the San Francisco Museum of Modern Art. Available from: http://www.businessinsider.com/google-search-event-live-2010-9.

Sangiovanni-Vincentelli, A., 2007: Quo vadis, SLD? Reasoning about the trends and challenges of system level design. *Proceedings of IEEE*, **95(3)**, 467–506. doi:10.1109/JPROC.2006.890107.

Searle, J., 1984: *Minds, Brains and Science*. Harvard University Press, Cambridge, MA.

Serres, M., 2001: *Hominescence*. Le Pommier, Paris.

—, 2003: *L'Incandescent*. Le Pommier, Paris.

Shannon, C. E., 1940: A symbolic analysis of relay and switching circuits. Report, Massachusetts Institute of Technology. Dept. of Electrical Engineering, Thesis (M.S.). Available from: http://hdl.handle.net/1721.1/11173.

—, 1948: A mathematical theory of communication. *ACM SIGMOBILE Mobile Computing and Communications Review*, **5(1)**, 3–55, Reprinted in 2001 with corrections from the *Bell System Technical Journal*, 1948. doi:10.1145/584091.584093.

Shimpi, A. L., 2013: The Haswell review: Intel core i7-4770K & i5-4670K tested. *AnandTech*, Retrieved March 18, 2016. Available from: http://www.anandtech.com/show/7003/the-haswell-review-intel-core-i74770k-i54560k-tested/6.

Simon, H. A., 1996: *The Sciences of the Artificial*. MIT Press, Cambridge, MA, 3rd ed.

Smolin, L., 2006: *Trouble with Physics: The Rise of String Theory, the Fall of a Science, and What Comes Next*. Houghton Mifflin Company, Boston and New York.

Smullyan, R. M., 1992: *Satan, Cantor & Infinity—Mind-Boggling Puzzles*. Alfred A. Knopf, New York.

Taleb, N. N., 2010: *The Black Swan*. Random House, New York.

The Edison Papers Project, 2016: The Thomas Edison papers. Rutgers University, retrieved March 10, 2016. Available from: http://edison.rutgers.edu/.

Wegner, P., 1997: Why interaction is more powerful than algorithms. *Communications of the ACM*, **40(5)**, 80–91. doi:10.1145/253769.253801.

Weizenbaum, J., 1966: ELIZA — a computer program for the study of natural language communication between man and machine. *Communications of the ACM*, **9(1)**, 36–45. doi:10.1145/365153.365168.

Wheeler, J. A., 1986: Hermann Weyl and the unity of knowledge. *American Scientist*, **74**, 366–375. Available from: http://www.weylmann.com/wheeler.pdf.

Wiener, N., 1948: *Cybernetics: Or Control and Communication in the Animal and the Machine*. Librairie Hermann & Cie, Paris, and MIT Press, Cambridge, MA.

Wolfram, S., 2002: *A New Kind of Science*. Wolfram Media, Inc., Champaign, IL.

Wolpert, D. H., 2008: Physical limits of inference. *Physica*, **237**(9), 1257–1281. doi:10.1016/j.physd.2008.03.040.

Wulf, A., 2015: *The Invention of Nature: Alexander von Humboldt's New World*. Alfred A. Knopf, New York.

Žižek, S., 2004: What Rumsfeld doesn't know that he knows about Abu Ghraib. *In These Times*, blog. Available from: http://www.lacan.com/zizekrumsfeld.htm.

Index

3Com, 105
3D Systems, 35
3D printer, 34

777, 118
787, 117
74181, 73

AAS, 110
Abelson, Hal, 76, 84
absolute zero, 133, 165
abstract model, 32, 37
Ada, 111, 246
Adams, Douglas, 84, 106
Adams, Scott, 30
Advanced Automation System, 110
AGI, 181
agile development, 111
AI, 115, 183
AI winter, 115
air traffic controller, 110
Airbus A350, 117, 220, 228, 247
AIS/Net 1000, 106
Akamai, 97
al-Khwarizmi, 145
aleph null, 157
Alexander, Christopher, 96
Algol, 89
algorithm, 145, 146, 173
almost surely, 232
alphabet, 187
ALU, 58, 69, 70
Amazon, 97, 114, 234
Andreessen, Marc, 92
Android, 91
Apache, 97
APL, 88, 96
app development platform, 113
App Store, 113
Apple, 90, 113, 234
Archimedes, 156
architecture, 78
Aristotle, 16, 33, 81
arithmetic logic unit, 58, 69
ARM, 81
Army, 111
artificial general intelligence, 181
artificial intelligence, x, 115, 183
assembler, 80
assembly language, 80
astrology, 21
AT&T, 106
autonomy, 247
Avatar, 241

aviation, 104
Avogadro's law, 55
axiom, 187

Babel fish, 106
backronym, 86
Backus, John W., 82
backward compatible, 78
BarCamp, 95
Bardeen, John, 14
Bayes' formula, 223
Bayes, Thomas, 220
Bayesians, 221
Bekenstein bound, 163, 186
Bekenstein, Jacob, 163
Bell, John Stewart, 216
bell curve, 230
Bell Labs, 49, 69, 86, 89, 91, 106, 128, 204
Bell System Technical Journal, 128
Bell's theorem, 216
Berkeley, 87
Berners-Lee, Tim, 5, 93, 103
Big Brother, 98
big data, 116
billiard balls, 209
bio-logic gates, 55
biology, 55
bipolar transistor, 15
bit, 69
black hole, 168, 172
black swan, xi, 20
Black Swan, The, xi, 6
Blackberry, 113
Boeing, 117, 118
Bohr, Niels, 10, 200
Bokor, Jeff, 50
Boltzmann constant, 133, 164
Boltzmann, Ludwig, 132, 164, 166
Boole, George, 68
Boolean algebra, 198
Boolean logic, 68
boot loader, 77
bootstrapping, 77
bot, 9
Boyle's law, 55, 56, 198, 202
Brattain, Walter, 14
Brin, Sergey, 184
Britannica, 8
Brooks, Fred, 78, 81, 88
Brooks' law, 88
Brown University, 146
browser wars, 92
BSD license, 95

digital machine, 73, 198
digital media, 101
digital physics, xiv, 143, 186, 208, 232
digital psyche, 183
digital rights management, 87
digital switch, 65
Dijkstra, Edsger, 75, 105, 119
Dilbert, 30
directory, 91
discrete event, 209
disorder, 134
distributed computing, 111
DNA, 180
dopants, 63
Dr. Seuss, 3, 225
drilling through a map, 41
drive-by-wire, 120
drone, 249
dualism, 237
Dunn, Stephen, 17
dynamics, 246
Dyson, George, 184

E-books, 114
eBay, 97
Echo, 234
Edinburgh, 84
effectively computable, 144, 149
Eich, Brendan, 92
Eindhoven University of Technology, 75
Einstein, Albert, 10, 21, 42, 102, 168, 199, 202, 216, 219
elastic, 210
ELIZA, 234
elizabot, 234
email, 110
emergent phenomena, 13, 57, 59
empiricism, 12, 13
encryption, 116, 202
Encyclopedia Britannica, 8
endure, 79
engineering, 7
ENIAC, 73
Enigma, 144
entanglement, 216
entropy, 132, 155, 165, 231
epiphenomenon, 57, 58
epsilon, 221
Ethernet, 105
Eudoxus, 156
eugenics, 15
event horizon, 168
Everett, Hugh, 200
evidence, 233
evolution, 19, 91, 180, 184, 245
evolutionary biology, 239
exabytes, 98
exclusive OR, 68
expected value, 231

FAA, 110
fab, 51, 63
fabless, 65
Facebook, 97, 117
Fairchild Semiconductor, 14, 104
falsifiability, 20, 238
Faraday, Michael, 43
Faraday's law, 177
Farber, David, 89
FCS, 111
Federal Aviation Administration, 110
feedback shift registers, 204
Fermilab, 168
FET, 14, 50, 63, 66
fidelity of a model, 41
field-effect transistor, 14, 63, 66
file system, 91
FinFET, 74, 113
Firefox, 92
flash memory, 91
fly-by-wire, 118
folder, 91
formal, 127
formal language, 186
Forms, 10
Forrester, 117
Fortran, 82
Foundation for Economic Education, 24
foundry, 65
Free Software Foundation, 87
free will, 196
freedom from choice, 54
frequentists, 221
Freudian unconscious, 28
functional languages, 84
functional programming, 83
Future Combat Systems, 111

Galapagos, 104
Gamma, Erich, 96
gang of four, 96
Gartner, 116
gas laws, 55
gate, 66
Gates, Robert, 112
Gauss, Carl Friedrich, 231
Gaussian distribution, 231
Gay-Lussac's law, 55
Geek Sublime, 53
General Accounting Office, 111, 243
general theory of relativity, 168, 202
genetics, 180
Ghemawat, Sanjay, 97
Gibbs, Josiah Willard, 132
Gibson, William, 121
Gill, Helen, 121
GLOBALFOUNDRIES, 65
GNU, 86

God, 7, 19, 164, 173, 184, 191, 199, 200, 219, 227
Gödel, Escher and Bach, 190
Gödel, Kurt, xv, 126, 160, 186, 201, 217, 219, 235, 237
Gödel's sentence, 189, 237
Golomb, Solomon Wolf, 41, 128, 204
Good, 10
Google, 91, 97, 104, 116, 184, 249
GPL, 87, 95
GPU, 51
gravitational waves, 42
Griswold, Ralph, 89

Hadoop, 97
halting, 146
halting problem, 146, 149, 152, 174, 190
hard disk, 91
hardware description languages, 62
Harris, Sam, 196
Haskell, 84
Haswell, 51
Hawking, Stephen, 168, 186, 190, 199, 201, 216, 219, 235, 237
health care, 117
Heffernan, Virginia, ix
Heisenberg, Werner, 10, 200
Hello Barbie, 234
Helm, Richard, 96
Henry, Joseph, 43
hidden variable, 216
hierarchical system, 75
Hilbert, David, 186
Hilbert's Program, 186
Hitchhiker's Guide to the Galaxy, 84, 106
Hofstadter, Douglas, 190
Hogan, Craig, 168
holographic principle, 172
Holometer, 168
Hopper, Grace, 89, 90
horse, 32
HP, 97
HTML, 93
Hu, Chenming, 50
human-computer interface, 244
Humboldtian model of higher education, 27
Hume, David, 227
hypercomputation, 152
hypertext, 6

IBM, 64, 78, 82, 97, 181
IBM Federal Systems, 110
IBM PC, 78
IBM System/360, 78
IEC, 108
IED, 249
imperative, 84
improvised explosive device, 249
incommensurability of paradigms, 31, 85, 191

incompleteness theorem, 186
indescribable, 192
inductance, 42
inductor, 42
inertia, 40
inference rule, 187
infinite, 153
information, 154
information technology, 47
information theory, 128
instruction set architecture, 78, 115, 118, 190, 198, 239
Intel, 51, 97, 104
Intel 8086, 78
intellectual property, 73, 108
intelligence, 247
interaction, 146
International Electrotechnical Commission, 108
Internet, 107, 109, 112, 184
Internet of Things, 99, 105
Internet Protocol, 109
interrupts, 119
intuitionistic logic, 188
inverter, 67
invisible hand, 25
iOS, 91
IoT, 105
IP, 73, 109
iPhone, 15, 16, 113, 245
IPv4, 112
IPv6, 112
iron wing, 119, 247
ISA, 78
ISO, 108
ISO/IEC JTC1, 108
it from bit, 172
ITU-T, 108
Ive, Jonathan, 102
Iverson, Kenneth, 88

Java, 89, 92, 97
JavaScript, 89, 92, 97
Jeopardy, 181
Jobs, Steve, 25
Jogalekar, Ashutosh, 22
Johnson, Ralph, 96
JPEG, 107, 108
jQuery, 93

Kalanick, Travis, 248
Kant, Immanuel, 29
Keynes, John Maynard, 226
King, Tsu-Jae, 50
known unknowns, 220
Knuth, Donald, 53
Kopetz, Hermann, 196
Kronecker, Leopold, 173

object-oriented design, 111
object-oriented language, 89
Occam's razor, 172
Ohm, Georg Simon, 38
one-to-one correspondence, 156
Oozie, 97
open source, 87, 90, 95
operating system, 77, 90
OR, 68
oracle, 153
OS X, 90
OSI model, 107
outsourcing, 101

P versus NP, 217
p-value, 222
paradigm, 28, 47, 239
paradigm shift, x, 29, 103
Pascal, 89
Patent and Trademark Office, 102, 109
patent holding company, 109
patent troll, 109
pencil, 24
Penrose, Roger, 169, 182
perishable insights, 117
philosopher, 10
PHP, 85, 97
phrenology, 21
physical layer, 107
physicalism, 237
Piccinini, Gualtiero, 169
Pig Latin, 97
Pitts, Walter H., 179
PL/I, 89
Planck, Max, 30
platform, 54
Plato, 6, 33
Platonic Good, 25, 244
Platonicity, xi, 6
Polonsky, Ivan, 89
polycrystalline silicon, 64
Pompidou Center, 102, 103
Popper, Karl, 12, 20, 29, 45, 162, 203, 216, 219, 238
portable, 89
positivist, 186, 237, 238
posterior probability, 223
presentation layer, 107
PRET project, 121
print-on-demand, 114
prior art, 109
prior probability, 223
privacy laws, 116
probability, 199, 203, 220
probability density, 135, 230
programming language, 81, 82, 198
Prolog, 89
proof, 188
prototype, 35

pseudocode, 150
pseudosciences, 21
pseudorandom bit sequences, 205
PTIDES project, 121
publish or perish, 243
punctuated equilibrium, 239
Putnam, Hilary, 179

quantum computing, 12, 166, 182
quantum gravity, 168
quantum mechanics, 134, 154, 167, 168, 197, 199, 202, 215, 216, 228, 239
QWERTY, 82

Rackspace, 97
Ramamoorthy, Chittoor, 82
RAND Corporation, 111
random chance, 219
Raphael, 33
Raytheon, 5
RCF, 191
Read, Leonard Edward, 24
Reagan, Ronald, 110
real closed fields, 191
real numbers, 156
realism, 163
reboot, 77
recursion, 86
Redford, James, 164
reductionism, 58
regime of applicability, 42
register transfer level, 73
relativity, 161, 168, 198, 199, 202, 214, 239
relay, 68
reliable transmission, 110
repeatability, 217
Research in Motion, 113
Resig, John, 93
reversible, 197
RIM, 113
RISC-V, 81
Ritchie, Dennis, 86, 91
rocket scientist, 18
RTL, 73
Rumsfeld, Donald, 28
Russell, Stuart, 249

Sackur-Tetrode equation, 165
Sanger, Larry, 4, 82
Sangiovanni-Vincentelli, Alberto, 54, 245
SARS, 185
School of Athens, 33
Schrödinger, Erwin, 10, 200
Schrödinger equation, 168, 197, 199, 200, 214, 215
Schrödinger's cat, 200
sciences of the artificial, 7, 45
sciences of the natural, 10, 45
Scotty, 244
scrambler, 204